Darwin
and the Art *of*
Botany

Darwin
and the Art *of*
Botany

OBSERVATIONS ON THE CURIOUS
WORLD OF PLANTS

....

James T. Costa and Bobbi Angell

WITH ARTWORK FROM OAK SPRING
GARDEN FOUNDATION

TIMBER PRESS
PORTLAND, OREGON

Photo and illustration credits appear on page 365.

Published in 2023 by Timber Press, Inc., a subsidiary of Workman Publishing Co.,
Inc., a subsidiary of Hachette Book Group, Inc.

1290 Avenue of the Americas
New York, NY 10104
timberpress.com
Printed in China

Text design by Rita Sowins / Sowins Design
Cover and jacket design by Vincent James

ISBN 978-1-64326-079-2
The publisher is not responsible for websites (or their content) that are not owned
by the publisher.

The Hachette Speakers Bureau provides a wide range of authors for speaking
events. To find out more, go to hachettespeakersbureau.com or email
HachetteSpeakers@hbgusa.com.
Catalog records for this book are available from the Library of Congress and the
British Library.

DEDICATED TO BOTANISTS THROUGHOUT THE WORLD
WHO CONTINUE TO STUDY PLANTS,
AND ARTISTS WHO CONTINUE TO DRAW THEM.

———— ······ ————

CONTENTS

FOREWORD

When Bobbi Angell first approached Oak Spring Garden Foundation with her concept for this book, we knew immediately that this would be an exciting way to bridge science and art, and also to share some of the treasures of Oak Spring Garden Library. The magnificent result more than justifies that early enthusiasm by adding visual stimulation to Charles Darwin's brilliant insights and carefully crafted prose.

Our benefactor Rachel Lambert "Bunny" Mellon once wrote, "The library comes first," and she set out a challenge: "How to share its beauty and knowledge in the most inspiring way giving human satisfaction and bringing books and nature together?" This book, created through the combined talents of James T. Costa and Bobbi Angell, provides a new and creative answer to Mrs. Mellon's rhetorical question.

Through her fascination with plants, Bunny Mellon was well aware of the impact and importance of Charles Darwin's work. As she created the library of Oak Spring Garden Foundation, she acquired copies of all Darwin's books. Aware of his deep botanical pedigree, she read *Phytologia: Or the Philosophy of Agriculture and Gardening* by his paternal grandfather, Erasmus Darwin, and she also purchased a copy of the first American edition of Erasmus's long discursive poem *The Botanic Garden*.

In *The Botanic Garden*, and especially in the section on *The Loves of the Plants*, Erasmus Darwin conveys his enthusiasm for the work of Carl von Linné, using anthropomorphic language to describe the parts of flowers and their function. Two generations later, it was his grandson, Charles, who developed the key concepts that underpin modern research on the form and function of plants. Charles Darwin's books on orchids, cross- and self- fertilization, and floral variation in the same species show us how floral diversity relates to the biology of plant reproduction. His insights reveal the meaning behind the exuberant variety of flowers. The varied illustrations in this book bring some of that floral diversity to life through works by some of the greatest botanical artists of all time, including many who were contemporaries of Erasmus and Charles Darwin.

From 1842, when he took up residence at Down House, up to his death in 1882, Charles Darwin devoted much of his life to his work on plants. He had begun to publish short observations on plants in The *Gardeners' Chronicle* in 1841. Subsequently, he published many articles on botanical topics, and six of his seventeen books, published between 1862 and 1888, are entirely devoted to plants. In addition to plant reproduction, Darwin also thought deeply about insectivorous plants, climbing plants, and plant movement. Plants feature prominently in his book on variation associated with "domestication." Tellingly, botanical examples are increasingly prominent in later editions of *On the Origin of Species*. Though his methodical experimental approach does not always make for easy reading, his insights shone light into areas of plant biology that had previously been dark.

Darwin's botanical interests were encouraged by his mentor at Cambridge, John Henslow, and also stimulated by his correspondence with the botanist Asa Gray at Harvard. However, his friendship with Sir Joseph Dalton Hooker, Director of the Royal Botanic Gardens, Kew from 1865 to 1885 and perhaps Darwin's closest scientific confidant, was especially important. Hooker had no hesitation at placing all of Kew's resources at Darwin's disposal. This was at a time when newly discovered plants were flowing into Kew from all over the world and often being cultivated, described, and illustrated for the first time. Consequently, Darwin had access to a vast range of botanical material, much of which is discussed and illustrated in this book. At Down House, Darwin grew plants obtained from Kew as well as from some of his many correspondents. Plants were perfect subjects for careful observation and experimentation. In a letter to Hooker in June 1857, he confessed that he found "any proposition more readily tested in botanical works ... than zoological."

It is interesting that one of the most perceptive observers of plants, and one of the greatest thinkers about nature, was neither an accomplished illustrator nor especially visual. For Charles Darwin, the primary focus was observations, experiments, and synthesis that led to new ideas. He once remarked, "How odd it is that anyone should not see that all observation must be for or against some view if it is to be of any service!" But documenting observations with illustrations was not a primary concern. The only illustration in *On the Origin of Species* is a simple diagram representing the divergence of species through geological time. Illustrations are also sparse in Darwin's other works, including those focused on plants. Bobbi Angell and James T. Costa have done a valuable service by connecting us to the variety of plants Darwin had in his mind and which featured in the development of his ideas. Their compilation of illustrations, selected from the collections of Oak Spring Garden Library, provide a unique perspective on Darwin's extraordinary capacity for synthesis. At the same time, they provide a new lens through which to highlight the quality, scope, and value of botanical art before and during Darwin's time. Bringing Darwin's text into conjunction with great botanical art is a masterstroke that adds interest and will bring new audiences to both.

It was an honor for Oak Spring Garden Foundation to partner with Jim and Bobbi in this exciting project and to share a sampling of the botanical art in Oak Spring Garden Library, which further celebrates the care with which Mrs. Mellon curated her unique collection.

SIR PETER CRANE FRS
President
Oak Spring Garden Foundation

PREFACE

This book is the product of a fortuitous meeting and collaborations made possible by two great institutions: The New York Botanical Garden's LuEsther T. Mertz Library and Oak Spring Garden Foundation, and to them the authors are deeply grateful. It all started with an idea that occurred to Bobbi Angell. Having drawn plants for botanists for more than forty years, Bobbi had long appreciated and been influenced by historical botanical art, admiring the rare books and artwork at The New York Botanical Garden (NYBG) and other institutions. A visit to Oak Spring Garden Foundation (OSGF) several years ago was a new and amazing treat, with Librarian Tony Willis giving her a grand tour of beautiful manuscripts in the collection. This refreshing view of paintings by both renowned and unknown artists inspired an idea of elaborating a historical text with historical art. Her immediate thought was to draw upon artwork from OSGF's incredible collections to illuminate climbing plants, more than 100 species of which are discussed in *On the Movements and Habits of Climbing Plants* by Charles Darwin, a book of painstaking research and insights that she felt deserved a display of beautiful contemporaneous artwork. Sir Peter Crane, President of Oak Spring Garden Foundation, and Tony Willis, Head Librarian, were enthusiastic from the first suggestion, and Bobbi was enthralled with the idea of sifting through books and manuscripts at OSGF to match up with all the plants Darwin studied.

Bobbi's idea led to an opportunity to meet Jim Costa at NYBG, where he had the good fortune to be immersed in the wonderful C. F. Cox Collection of Darwiniana in the Mertz Library as a Mellon Visiting Scholar. They met for coffee one fine day, and Bobbi bounced the idea: a book on Darwin's fascinating work with climbing plants, illustrated with beautiful and historically significant botanical artwork. Jim thought it was novel and exciting, as any fellow self-confessed Darwin and plant nerd would agree—what a neat way to spotlight Darwin's fascinating and creative experimental work with climbers as well as illustrate the diversity and beauty of these plants as seen through artists' eyes.

Together we proposed the idea to Tom Fischer at Timber Press, who enthusiastically suggested we expand our ideas, focusing not just on *Climbing Plants*, but all six of Darwin's botanical books—a prospect exciting and daunting in equal measure. Bobbi then had the challenging task of picking and choosing among the many wonderful plants studied by Darwin that were also represented in Bunny Mellon's fabulous botanical art collection, ultimately selecting forty-five species that represent the range of Darwin's botanical investigations, with selected passages from his writings that exemplify his working method and insights. Jim took the lead on the general introduction and writing accounts for each plant, highlighting Darwin's interest and methods in a behind-the-scenes kind of way, and aspects of the plant's natural history, ecology, and evolution more generally.

Although we both began this project as plant lovers, Darwin taught us much more about plants than we could have imagined, inspiring a deeper appreciation for everything from mechanistic structure, function, and physiology to adaptation and variation in an evolutionary context. Delving into Darwin's working method was equally illuminating. We hope that the marriage of art and science that we present here will inspire readers to see plants—too often considered mere background or admired solely for the beauty of flower or foliage—in a new light altogether: beautiful, yes, but also as exquisitely adapted organisms with a rich evolutionary history. We hope, too, that we will help inspire a new appreciation for beautiful botanical art and beautiful science, and the creative eye and spirit that underlie both.

INTRODUCTION

Charles Darwin was quite an artist. No, not that kind—he appreciated the visual arts but was hopeless with a sketchpad, let alone a canvas—his aptitude for drawing was perhaps second-to-last in any ranking of his abilities (his handwriting, notoriously illegible, no doubt coming in at the bottom). But Darwin was surely an artist of science—an artful asker of questions and crafter of studies both observational and experimental. There is, after all, an art to science—beauty in theory, elegance in experimental design, a satisfying harmony in the consilient resonance of supporting evidence. Of course, Darwin would not have described his work as "beautiful" or "elegant"—in fact, just the opposite. "I love fools' experiments," he quipped to one visitor. "I am always making them."[1] He liked to poke fun at himself, but there is no question that the self-described "experimentiser" was a master of composition and had a deft touch when it came to those "fools' experiments"—myriad curious research projects unfolding everywhere from his gardens and greenhouses to the meadows and woodlands around Down House, his home of forty years just south of London. That home was a veritable research station, with Darwin as chief scientist and his wife, children, extended family, servants, and friends his ever-obliging research assistants. Many a correspondent helped too, some recruited through Darwin's many "crowd sourcing" letters to magazines and newspapers.

Darwin is best known for his masterwork *On the Origin of Species*, published in 1859, a book that he described as "one long argument." What he meant by that is most evident in the topics treated in the second half of the book, with chapters dedicated to a diversity of

Photo of Charles Darwin at age seventy-two on the veranda at Down House, with
Parthenocissus twining up the post. Photo taken in 1881 by Messers. Elliot and Fry,
published in *The Life and Letters of Charles Darwin* (1887).

subjects (behavior, fossils, geographical distribution, comparative anatomy, and more), lines of evidence that, he argued, were collectively united and explained by his theory of evolution by natural selection. The fame of *Origin* tends to overshadow the fact that Darwin followed it with some dozen more books, beginning with his curious volume on orchid pollination mechanisms in 1862. How odd, some thought, to follow so sweeping a book as *Origin* with one so narrowly focused. But there was method to the madness: the orchid book signaled the direction his work would take for the next twenty years—a corpus of work that was *Origin* writ large, myriad subjects with an eye to extending and reinforcing the explanatory power of his theory. It was, collectively, one longer argument. The orchid book also signaled a new favorite research subject: plants. Half of his post-*Origin* books treated botanical subjects—climbing plants following orchids, then carnivorous plants, forms of flowers, pollination, and more. Darwin didn't consider himself a competent botanist, but his friends knew better—Asa Gray, at Harvard, applauded and encouraged his botanical investigations, writing, "it will be fruitful in your hands."[2] The metaphor was more appropriate than Gray may have anticipated; Darwin's botanical investigations grew, blossomed, and bore rich fruit in the form of remarkable new discoveries, launching whole new disciplines that have themselves grown and continue to bear fruit today. For a self-described amateur when it came to botany, Darwin sure had a green thumb when it came to growing both a garden and ideas.

Some might say he came by these talents honestly. Charles Darwin was born on 12 February 1809, the second son of physician Robert Darwin and Susannah Wedgwood. The families on both sides were accomplished: Robert was the son of the famed physician and poet Erasmus Darwin, friend to a "who's who" of Enlightenment Britain's intellectual circles, whose best-selling poetry extolled scientific subjects, including the famous *The Loves of the Plants*—an at-times-titillating exposition of Linnaeus's sexual system of botany.

Susannah was the daughter of Josiah Wedgwood, founder of the celebrated pottery works and prominent member of those intellectual circles frequented by Erasmus—they were friends with and founding members of the Lunar Society, that remarkable circle that included among others James Watt (of light bulb wattage), industrialist Matthew Boulton, pioneering chemist Joseph Priestley (discoverer of the element oxygen), and corresponding member Benjamin Franklin, the celebrated American statesman and inventor. An appreciation of flowers and gardens was very much in the tradition of both families. Charles Darwin's childhood home, The Mount, in the bustling market town of Shrewsbury, featured a bounteous hillside garden, and the gardens and pleasure grounds of nearby Maer, the Wedgwood estate, were originally laid out by Capability Brown. Darwin's cousin and future wife Emma Wedgwood, grew up there, reveling in those gardens; he used to say that Emma only cared for flowers that were grown at Maer.[3]

But it's fair to say that, for Charles Darwin growing up, flowers and gardens were more backdrop, a theater for childhood romps, than interest per se, and as an affable teen he was mostly into riding his horses and hunting with his dogs (though there was the smelly chemistry lab he set up in the stables with his older brother 'Ras). As for future profession, there was no question: in the family tradition he would become a doctor; he was duly packed off to Edinburgh at age sixteen to join 'Ras in medical school. That didn't work out too well. After witnessing a couple of horrific operations, one on a child, he quickly discovered that he couldn't stomach the blood and screaming. He left Edinburgh after two years, heading to Cambridge where he was expected to prepare for a career in the church (country parson was a perfectly respectable profession for someone of his social class). Edinburgh had not been a total bust, however. In his time there, he studied zoology with Robert Edmond Grant, who introduced him to the fascinating world of marine invertebrates (on which Darwin conducted his first research project) as well as the somewhat disreputable world of unorthodox theorizing. Grant had studied in Paris and was a devotee of French savant Jean-Baptiste de Lamarck, who had developed an elaborate theory of species change, or "transmutation." Such unorthodox ideas, deemed contrary to scripture, had only been tolerated in Britain since the bitter aftermath of the terror in France, when transmutation became synonymous with revolution, even sedition. The young Charles Darwin was no wild-eyed revolutionary—he had no interest in such scandalous ideas and shrugged off his professor's transmutationist musings. Yet a seed was planted.

Moving on to Cambridge, he commenced training for the church. But that was not to be either, thanks largely to beetles and botany. On the beetling front, Darwin was caught up in collecting, then all the rage among the undergraduates (puzzling as that may be to modern readers!). A taste for natural history was not contrary to the country vicar's life—just the opposite, as Britain had (and has) a great tradition of talented and accomplished vicar-naturalists, and Darwin came under the influence of one such, the kindly Reverend John Stevens Henslow, professor of botany. A clergyman like all Cambridge faculty, Henslow led legendary botanizing rambles after church each Sunday. Darwin was often in attendance and more and more taken with his professor and the subject, becoming known as "the man who walks with Henslow." He was excited by geology too, an up-and-coming science offering astounding new insights into earth's antiquity—Darwin found his professor, Reverend Adam Sedgwick, a "capital hand" at drawing "large cheques on the bank of time."[4] As his interest in natural history grew steadily stronger, Darwin's fate may have been sealed with his reading of the great Alexander von Humboldt, the German polymath whose electrifying *Personal Narrative* chronicled his explorations in Mexico and northern South America and inspired a generation of naturalists to travel, observe, and study nature holistically. Darwin started scheming, hoping to convince a band of friends to travel with him to the Canary Islands to follow in Humboldt's footsteps. That trip ended up fizzling, but Darwin's disappointment evaporated when another, even more

exciting prospect was presented thanks to Henslow: a voyage around
the world on the Royal Navy surveying ship HMS *Beagle*, as naturalist
and gentleman-companion to the captain, Robert Fitzroy.

The only fly in the ointment was his father. Robert Darwin was
less than pleased with this "wild scheme." Wanting his son to take
holy orders and settle down already, he put his foot down—this was
a waste of time, and possibly damaging to Charles's future prospects
in the church. Evidently however, he gave a caveat: if any respectable
person with good sense could give him reasons why his son should go
on such a voyage, he might reconsider. A despondent Darwin saddled
up his horse and headed to Maer, where his uncle, Josiah Wedgwood
II, lent a sympathetic ear—and proved to be just the respectable and
sensible person needed, helping Darwin make a case to his father.[5]
It worked, and over the next five years, Charles Darwin had the ride
of his life, making extensive geological observations (thanks to a
crash-course in field geology with Sedgwick) and collecting fossils,
zoological specimens, and lots and lots of plants.

The animals Darwin encountered get most of the attention—
fascinating finches and charismatic mega-fauna like the giant
tortoises of Galápagos and even larger extinct beasts of the South
American mainland, including great armadillo-like *Glyptodon*
and the elephant-sized ungulate *Toxodon*. But Darwin himself was
about as taken with the plants. Musing over the birds of Galápagos,
Darwin noted in his journal that "I certainly recognise S. America"
in their affinities but also wondered "would a botanist?"[6] And later,
again musing on the significance of Galápagos species, he declared
that "the botany of this group is fully as interesting as the zoology."[7]
On his return, Darwin hoped that his old mentor Henslow would
analyze his South American plants. Up to his neck in work, however,
Henslow didn't have time and ultimately passed them on to a young
up-and-coming botanist named Joseph Hooker at the Royal Botanic
Gardens, Kew. Darwin had by then been converted to the heretical notion of transmutation that his Edinburgh professor Grant

Down House, Darwin's home in Kent. Wood engraving by J. R. Brown,
courtesy of the Wellcome Collection

had spouted, struck by both the Galápagos birds and those amazing
South American fossils he had collected. Together these presented
unmistakable patterns of species' relationships in space and time—
the only explanation, he realized, was the possibility that species
can change, giving rise to new species over time. Keen to see if the
plants he collected told the same story as the birds, he commented
to Hooker, "I hope the Galapagos plants ... will turn out more
interesting than you expect. Pray be careful, to observe, if I ever
mark the individual Island of the Galapagos islands, for the reasons
you will see in my Journal." Hooker confirmed that the botany did
indeed provide an analogous case, exclaiming that there was an
"xtraordinary difference between the plants of the separate Islands
... a most strange fact."[8] Darwin knew why but didn't clue Hooker
in on his transmutational thinking—not yet. But the two were to
become close friends, and Darwin did eventually take the botanist into his confidence. Unlike Grant (and his own grandfather),

Darwin was not a devotee of Lamarck. In October 1838, about a year and a half after becoming a closet transmutationist, he had a eureka moment that brought him to the principle of natural selection. Evolutionary change, he realized, was not so much a linear succession of species, but rather like a branching and rebranching tree—an appropriate metaphor given the importance of botany in his thinking.

From the get-go, botany was an important part of Darwin's growing database, providing lines of evidence in support of his still-secret theory of "descent with modification," as he described it, by natural selection. He and Emma Wedgwood were married in 1839, and after a two-year residence in London, the young couple moved to the country. They found just the place in the village of Downe, a handsome home called Down House, with plenty of room for a growing family—and growing gardens—surrounded by fields, meadows, and woodlands. All of these were sites for Darwin's botanical investigations: plant geography, ecology, hybridization, pollination, and more were on the research agenda. He even got into field botany, identifying plants with the help of Catherine Thorley, the children's governess. In 1855, he wrote Hooker celebrating his successes, "I have just made out my first grass, hurrah! hurrah!" and admitting, "I never expected to make out a grass in all my life."[9]

It was in those years, especially, through the 1850s leading up to *Origin*, that Darwin first *really* got into doing his "fools' experiments"—curious and often quirky studies that might have seemed to the casual observer like the random enthusiasms of a doddering naturalist. Once, he plucked an artificial flower from Emma's bonnet and planted it in the garden to see if bees paid any attention to it. He also dunked ducks in tanks of duckweed to see if the diminutive plant really would stick to their backs, netted clovers to keep bees from getting to them to compare the resulting seeds with those visited by bees, minutely mapped the pollinator "gangways" of flowers, and put seeds in jars of saltwater to simulate floating at sea, periodically planting batches of them to gauge how long they could survive (and therefore how far they could be carried by currents, to populate remote islands). Then there was that memorable time he tried to "feed" clippings of his toenails to sundews. Odd and amusing, yes, but also highly informative.*

Darwin had by this time taken Hooker into his confidence, followed by two other friends, geologist Charles Lyell in Britain and botanist Asa Gray in America. He had started to pull it all together into a book, a big book that he called *Natural Selection*. But he was taking his time and might have continued indefinitely working on it and collecting evidence to back up the theory had he not been scooped by naturalist and fellow Briton Alfred Russel Wallace. Thirteen years younger than Darwin, Wallace was then collecting and exploring in the Malay Archipelago (mainly modern Indonesia), where he had been hard at work trying to solve the mystery of the origin of species himself. He had his own eureka moment in February 1858, discovering natural selection just a few years after his arrival in the east. By a quirk of fate, he had been casually corresponding with Darwin, and each knew that the other was interested, in a general way, in the nature of species and varieties. Wallace wrote out his newly discovered theory and mailed it off to Darwin, of all people, who (needless to say) was broadsided and devastated. Darwin's friends came to the rescue. Hooker and Lyell quickly arranged to present some of Darwin's unpublished writings on the subject together with Wallace's paper. Darwin wrote Wallace, explaining that he had been working on just that theory for nearly twenty years. Now he was under the gun to come out with his book and prove his priority. But his big book was *too* big and would take way too long to finish. He ultimately

---

* To see just how informative—and try your hand at DIY versions of some of Darwin's experiments—see Costa 2017.

condensed his original book down to an "abstract" of nearly 500 pages (!) and published his "one long argument" as *On the Origin of Species*. Wallace was taken with the book, deeply impressed and exceedingly gracious about the turn of events.*

After publication, Darwin forged on ahead, continuing work on subjects that further illuminated the principles laid out in *Origin*. He decided to start with orchids for three reasons.[10] First, he had already been studying them and found their intricate structural adaptations for pollination just astounding, "as varied and almost as perfect as any of the most beautiful adaptations in the animal kingdom." He had written Hooker in a state of orchidelirium, making just this point: "Have pity on me and let me write once again on Orchids for I am in a transport of admiration at most simple contrivance, and which I should so like you to admire. How I wish I was a Botanist."[11] (As if!) Second, he said, the exquisite adaptations of orchids function to ensure cross-pollination and do so with amazing precision, under-scoring a point he made in *Origin* that at least occasional crossing was practically a universal law of nature. Finally, some had criticized *Origin* for lacking the reams of supporting evidence that such a startling theory demanded. Darwin had stressed that the book was an abstract, after all, but the criticism still stung, and he intended to prove that he did indeed have more evidence, stating, "I wish to show that I have not spoken without having gone into details."

But there was more to Darwin's orchid research than quieting his critics. Though they achieve the same end—enforcing cross-pollination by affixing pollen packets, called *pollinia*, to insect couriers—the

---

* Some authors have portrayed Darwin as having wronged Wallace in this episode, charging that Darwin allowed Charles Lyell and Joseph Hooker to present his own unpublished writings with Wallace's paper to ensure priority, or, worse, that, in the process, he pilfered ideas from Wallace to complete his own theory. It is reasonable to debate the former claim, but there is absolutely no evidence to support the latter. For analysis and discussion of the evidence, see Costa 2014.

Darwin's study at Down House, complete with experimental plants.
Wood engraving by A. Haig, courtesy of the Wellcome Collection.

intricate adaptations of orchids are not all the same. Different orchid
groups accomplish pollination in different ways, and there was no
single "perfect" adaptation that a creator may have made; rather, they
represented variations on a theme—more consistent with the vagaries
of evolutionary change than with divine design. Asa Gray was supportive and began to understand Darwin's goal. "Of all the carpenters for
knocking the right nail on the head," Darwin wrote admiringly to
Gray, "you are the very best: no one else has perceived that my chief
interest in my orchid book, has been that it was a 'flank movement'
on the enemy. ... It bears on design—that endless question."[12]

   In fact, all of Darwin's botanical works were "flank movements"—
they were all of a piece: demonstrations, applications, and extensions
of his theory designed to back up and reinforce his arguments for
evolution by natural selection. And along the way, he made more than a

few contributions to the understanding of plant biology. He demonstrated the digestive physiology of carnivorous plants, for example, elucidated the function of heterostyly (flower morphs with different stamen and pistil lengths), discovered "circumnutation" (a term he coined for the repetitive circular motion of growing shoots and searching tendrils), mapped the intricate pollination mechanisms of orchids and other flowers, and on and on. This is not to say that Darwin always got it right in the modern view—such is the nature of scientific inquiry that he sometimes barked up the wrong tree, like when his tendency to emphasize the animal-like qualities of carnivorous plants led him to believe that these plants have a nervous system of sorts. Similarly, his pioneering research into plant movement could only go so far without an understanding of plant hormones, not discovered until much later by plant physiologists. At a time when laboratory science was becoming highly professionalized, Darwin was derided by some scientists for his amateurish "country house" experiments, which lacked the controlled conditions and precision instruments of a proper lab. Sure, there is some truth to that charge, but there is also no denying that Darwin made discoveries through his home-spun experiments that proved foundational to modern studies of plant physiology, ecology, and evolution.

Darwin didn't do it alone, of course, as he readily acknowledged. First, he knew he stood on the shoulders of such botanical giants as Alexander von Humboldt, Augustin de Candolle, Carl Linnaeus, Konrad Sprengel, Karl Friedrich von Gartner, Sir Robert Schomburgk, Robert Brown, John Stevens Henslow, and others. What's more, he appreciated the assistance he received from all quarters, helpers whose efforts nurtured his gardens, literal and intellectual—the botanical beauties of Down House gardens and greenhouse as well as his verdant garden of ideas. Assistance came from family, friends, and correspondents far and wide who willingly collected specimens, made observations, and lent their expertise. At home, his wife, Emma, and their seven children were ever-ready research assistants, even well

Darwin in his Greenhouse. *Illustrated London
News*, December 1887.

into adulthood. Son Leonard recalled how their investigations had the character of both "a game of play and of a scientific inquiry; and in so far as we were at play, my father was like a boy amongst other boys."[13] Darwin's sister-in-law Sarah Wedgwood collected plants for him, and his Wedgwood nieces eagerly made botanical observations while on holiday, writing to him, "Dear Uncle Charles, of 256 specimens of *Lythrum* gathered this morning from different plants, we find 94 with long pistil, 95—middle length pistil, 69—shortest pistil."

He was always effusive in his thanks. "My dear Angels!" came his swift reply; "I can call you nothing else ... the enumeration will be invaluable."[14]

It helped to have well-connected friends, too. Botanist Joseph Hooker, first as Assistant Director and then Director of the Royal Botanic Gardens, Kew, supplied Darwin with all manner of exotic plants from around the world for his experiments, and Daniel Oliver, Keeper of Kew's Herbarium, James Veitch, proprietor of the Royal Exotic Nursery in Chelsea, orchidologist extraordinaire James Bateman, and botanist Alexander G. More on the Isle of Wight, among others, all helped Darwin with specimens or observations or both. John Lindley, editor of the popular *Gardeners' Chronicle*, was always glad to publish Darwin's articles as well as frequent letters requesting the help of readers with this or that observation or experiment—Darwin was the original crowdsourcer. Assistance also came from far afield—the leading US botanist Asa Gray was a friend and frequent correspondent, and Gray put Darwin in touch with able observers in the American field: businessman and botanist William Marriott Canby in Wilmington, Delaware, and the talented writer and naturalist Mary Treat in Vineland, New Jersey, both of whom are cited extensively in Darwin's book on carnivorous plants. In other quarters of the world, Italian botanist Federico Delpino lent a hand, as did Roland Trimen in South Africa and German expatriates Fritz Müller in Brazil and Hermann Crüger in Trinidad. You get the picture—Darwin's working method shows that science is a collaborative enterprise.

Pl. 2.

Mfs Drake del.   M. Gauci lith. P. Hirck, Vincent Bedford Sq.e

CATASETUM MACULATUM.

Publd by J. Ridgway & Sons 169 Piccadilly. July 1.1837.

Catasetum maculatum. Lithograph, drawn by Sarah Anne Drake, for *The Orchidaceae of Mexico and Guatemala*, by James Bateman.

......

Darwin's research on plants led to six books entirely dedicated to botanical subjects, along with some seventy-five papers. His botanical books include *The Various Contrivances by Which Orchids are Fertilised by Insects* (1862, 1877), *On the Movements and Habits of Climbing Plants* (1865, 1875) *Insectivorous Plants* (1875, 1888), *The Effects of Cross and Self-Fertilisation in the Vegetable Kingdom* (1876), *The Different Forms of Flowers on Plants of the Same Species*, (1877) and *The Power of Movement in Plants* (1880). He also discussed plants in *On the Origin of Species* (6 editions, 1859–1872) and *The Variation of Animals and Plants Under Domestication* (1868, 1875). His contributions to the *Gardeners' Chronicle* began in 1841 and continued for thirty-six years, both queries and articles, in addition to articles on botanical subjects in the *Journal of the Linnean Society of London*, the *Journal of Horticulture*, and the *Annals and Magazine of Natural History*.

Darwin covered an astonishing diversity of plants in his far-ranging studies: 125 species of climbers, nearly 70 orchid genera, including native and tropical species, 20 carnivorous plants, more than 200 species that he experimented with for pollination and movement, and dozens of fruits and vegetables. Perusing his books, it's hard not to be inspired to grow and observe these same plants—easy to do since so many are staples of the kitchen garden or favorite ornamentals. Watching pollinators land on flax flowers; noting the age sequence of foxglove flowers; growing common and unusual vines such as hops and passion flowers to watch how they twist and twine on supports; noticing how peanuts bury themselves in the ground as they mature, as do the developing fruits of cyclamen, twisting and turning downwards toward the soil; watching three-leaf clovers fold up on a cool night; noting the slow-motion reach of sundew "tentacles"— these observations can be made anywhere, á la Darwin, out on a walk, in the garden, on the trail, or amid the houseplants.

Darwin's writing is detailed and extensive, perhaps a little *too* detailed in places, which is why we selected excerpts. There is no denying that even these can be a bit dense—but it is well worth braving his prose to be rewarded with a very good sense of his painstaking technique, those minutiae of careful observation and experimentation that lie at the heart of good science. We can credit Darwin for helping us decide which excerpts to select—he was often a bit self-deprecating and at times frankly acknowledged his overly detailed writing. For example, in sending Gray the page proofs of *The Effects of Cross and Self Fertilisation in the Vegetable Kingdom* for review, he included a heads-up that the text was on the dry side: "Please observe that the 6 first chapters are not readable, and the 6 last very dull."[15] Not a ringing self-endorsement. "Still," he said, "I believe the results are valuable."

In this volume, we present forty-five plants studied by Darwin, representing the broad range of his extensive botanical research, with enough detail in the excerpts to give an appreciation for his working method and the remarkable depth of his investigations. Collectively, our selections cover Darwin's major botanical research threads:

ORCHIDS Although Darwin had started observing native orchids nearly twenty years earlier, it was in the spring of 1860 that he first probed two native *Orchis* species with the tip of a pencil to simulate the visit of an insect pollinator; the pencil played the role of a moth's probing proboscis, neatly extracting the pollen packets by their sticky bases. Intrigued, he soon started acquiring exotic orchids from the Royal Botanic Gardens, Kew and Veitch's nursery, and pickled flowers from Trinidad. His interest in orchids was deeper than pollination—the complex structure and intimate pollinator relationship of these remarkable flowers provided a case study in exquisite adaptation through natural selection, as different orchid groups (lineages, to him) have different floral structures modified for the same function: the kind of variations-on-a-theme pattern that spoke more of an evolutionary history than divine design to Darwin.[16]

CROSS AND SELF-FERTILIZATION, VARIATION, POLLINATION, AND FORMS OF FLOWERS Darwin's interest in cross-pollination went way back to his earliest speculations about transmutation— what would become known as evolutionary change. Heritable variation was a key ingredient, through natural selection, but where did variation come from? He didn't know, but it seemed to him that crossing, or out-breeding, was crucial for some reason. Flowering plants, in all of their wonderful diversity, became the organisms of choice for studying this question. Literally rooted to the spot, how did they find mates? They didn't, instead enticing insect go-betweens to do the matchmaking for them. Today we take for granted the role of insects in pollination, but not so in Darwin's day; the conventional view was that insects had little to nothing to do with pollination and fruiting. Darwin championed the work of German naturalist Christian Konrad Sprengel, who argued for the vital role of insects in pollination in his beautiful 1793 book *Das entdeckte Geheimnis der Natur im Bau und in der Befruchtung der Blumen* (*The Secret of Nature in the Form and Fertilisation of Flowers Discovered*). Scrutinizing flowers became a passion for Darwin, dissecting them, mapping out how insects enter and leave, noting how and where pollen is deposited, and delighting in "irritable" flower structures like the trigger-sensitive stamens of barberry or orchids that can fire off their pollen packets ballistically. He reveled in meticulously documenting the adaptive structure and function of flower after flower, studies complemented by equally meticulous controlled crosses over multiple generations to document the beneficial effects of outcrossing vs. self-fertilization. While he could not know the ultimate genetic basis of variation and how it is generated, he hypothesized that crossing in plants, and sexual reproduction gen- erally, is all about bringing together heritable variation in endless, myriad, combinations—beneficial for the individual constitution of offspring, and providing raw material for natural selection to act upon.

In the course of his floral studies Darwin stumbled upon a curious phenomenon: short- and long-style morphs of primrose flowers, termed "heterostyly." His Cambridge mentor John Stevens Henlow had reported these primrose flower morphs years earlier but had left it at that. Darwin became intrigued, and he performed extensive experimental crosses between and within morphs to figure out their function. After testing and rejecting his favored hypothesis, his eventual conclusion, which proved correct, was that heterostyly is an adaptation to promote outcrossing. Darwin had studied dozens of heterostyled plants, performing thousands of crosses by hand with dimorphic and trimorphic species. His data showed that between-morph pollination yields a greater abundance of fruits and seeds than within-morph crosses. Later, in his autobiography, Darwin commented that "no little discovery of mine ever gave me so much pleasure as making out the meaning of heterostyled flowers."[17]

CLIMBING PLANTS The earliest record of Darwin noticing climbers comes from his time in Brazil while traveling aboard HMS *Beagle*, where he was impressed by the abundance of "twiners entwining twiners." Many years later, he read an 1858 paper about vines written by Asa Gray at Harvard and asked Gray to send him seeds so he could make his own observations. Darwin had been under the weather and needed something relatively easy to study; climbers became just the thing. He grew Gray's *Echinocystis* seeds and followed up with one species after another, requesting loads of plants from Kew. " I am getting very much amused by my tendrils – It is just the sort of niggling work which suits me" he wrote to Hooker. By 1863, he was "madder than ever on tendrils," and wrote rhapsodically to Gray that tendril diversity "is beautiful in all its modifications as anything in orchids."[18] In 1865, Darwin reported his observations and experiments with some twenty leaf-climbing and tendril-bearing taxa in a lengthy paper to the Linnean Society of London—subsequently expanded to *The Movement and Habits of*

Climbing Plants (1875). He sought in climbers the same patterns he had seen in orchids: exquisite adaptation that sheds light on evolutionary history. As he had with orchids, Darwin traced out structural "variations on a theme" in climbers, structures with a function in some groups adapted for a different function in others. He recognized five categories of climbers: *hook and root* climbers, which he had minimal interest in; *twiners*, the most "primitive" climbers; *leaf climbers*, with various parts of the leaves enabling the plants to scramble up supports; and *tendril-bearers*, with their remarkable climbing abilities described in great detail. He documented circular motions traced by twiners and tendrils, noting the timing of their movement by using a visual marker to track their progress—work that led him to coin the term "circumnutation" for the elliptical movement of plant tendrils and growing shoots. Beyond adaptive structure and function, Darwin was struck by the animal-like qualities of climbers—he marveled at their touch sensitivity, and clear ability to "see" light and shadow and move accordingly. The sensory perception of climbing plants was yet another evolutionary lesson for Darwin: a fundamental link with animals, he was sure.[19]

INSECTIVOROUS PLANTS Poorly understood in Darwin's day, carnivorous plants are the subject of his third botanical book. His investigations started when he came across sundews by chance while on summer holiday near the coast and he carried a few back to the family's rental cottage for closer scrutiny. His attention snared as surely as an insect, he started growing sundews in his greenhouses, experimenting with simple techniques such as touching the dew-tipped "tentacles" on the leaves with a camel-hair brush and sprinkling countless tidbits onto the leaves to see what they would and would not respond to. Like climbers, here was another group with clear animal-like qualities. He carried out extensive studies of their "behavior," assisted by observers abroad, including naturalist Mary Treat in New Jersey, who sent detailed observations of sundews and Venus fly traps and helped conduct experiments on the dietary preferences of these most animal-like of plants. His fascination with those qualities, not least the astonishingly fast snap-trap action of Venus fly traps—"the most wonderful [plant] in the world"[20]—led Darwin to team up with physiologist John Burdon-Sanderson to investigate if these remarkable plants had a nervous system of sorts.

PLANT MOVEMENT Convinced as he was of common ancestry of plants and animals deep in evolutionary time, Darwin became ever more interested in the other seemingly animal-like qualities of certain plants. His work on climbers grew into an interest in plant movements in general, and he and his son Francis launched into a series of investigations on mechanisms of movement in various plant parts—shoots, stems, leaves, or tendrils—and their context, for example, movement in response to gravity (gravitropism) or light (phototropism) and the nocturnal "sleep" movement of shoots and leaves (nyctinasty). Here was yet another case study in evolutionary variations-on-a-theme; all these forms of movement, Darwin realized, represent modifications of the slow, circular movement, circumnutation, found in all growing shoots. These studies inspired pioneering experiments on the sensory perception of embryonic shoots (the coleoptile) and roots (the radicle) by the father and son Darwin team, anticipating the discovery of plant-growth hormones and cell signaling decades later.

The movement experiments gave Darwin an even deeper appreciation for the animal–plant link, and he declared "there is no structure in plants more wonderful ... than the tip of the radicle," comparing it to a brain from "one of the lower animals."[21] Over a lifetime of study in his literal and figurative garden of botanical marvels, each of Darwin's discoveries thrilled and delighted him in turn, each exciting wonder and inspiring his admiration to ever-greater heights. It is no surprise that he should find this, the last of his major botanical investigations, to be the most amazing yet.

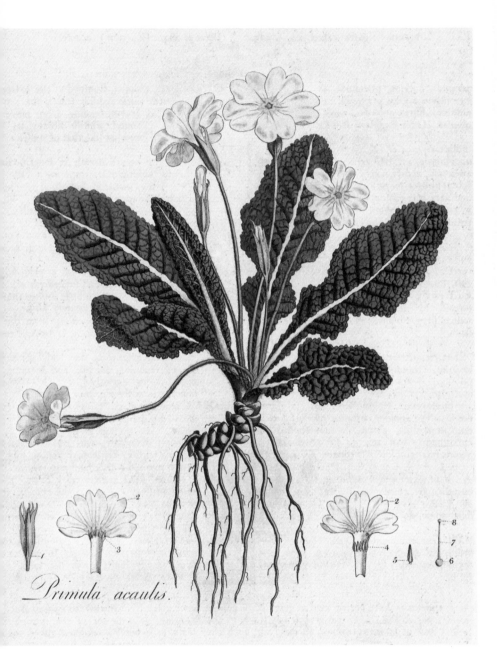

Primula acaulis.

Primula acaulis. Hand-colored engraved plate from *Flora Londinensis*,
published by William Curtis.

Ipomoea nil. Bodycolor on vellum, painted by Jan Withoos, *Dutch Florilegium.*

......

Although Darwin lived in a golden age of botanical art and illustration, with paintings and engravings adorning many a book—the more lavish ones supported by subscribers and wealthy patrons—he did not make use of the fine work done by the celebrated artists of the day in his own books. Rather, Darwin preferred detailed woodcut illustrations for his botanical works, beginning with remarkable illustrations of dissected flowers for his orchid book by George B. Sowerby (who then taught Darwin's sons George and Francis how to draw them) carved into wood-cuts by Mr. Cooper. Darwin was probably not interested in attractive illustrations for his books—most of the time not even showing the whole plant, or even entire flowers—because he saw them as narrowly focused scientific tomes. We seek to remedy this deficiency.

Botanical illustration is an exquisite synergy of artistic and scientific practice, in which plants are depicted not only with great accuracy but also with a refined aesthetic that captures their beauty. Flowers and plants have always found a place in art from civilizations all around the world, but in Europe, classical knowledge was revived during the Renaissance with a new focus on the medicinal properties of plants—in the late fifteenth and early sixteenth centuries, previously symbolic representations of plants in herbals and books of hours were transformed into elegant lifelike reality by artists such as Leonardo DaVinci (1452–1519) and Albrecht Dürer (1471–1528).

In the seventeenth century, new interest in horticulture and the uses of plants stimulated further refinement of botanical art. Explorers encountered many species that have commercial value in horticulture, and plant hunters were dispatched around the world by enlightened and wealthy patrons to bring home exotics to enrich their gardens and landscapes. Arrivals in Europe of exotic plant species from Africa, the Far East and the New World, especially to Spain, Holland, and France, created new demand: discoveries and beauty needed to be captured through visual representation. Introductions of new plants gathered even more momentum in the eighteenth and nineteenth centuries with Britain's expansion of its colonial territories.

The explosion of botanical exploration through the centuries brought opportunities for artists to capture the form and beauty of new discoveries in florilegia and scientific treatises. Multiple-volume journals and large format books were produced, illustrated by legions of exceptional artists and made available by talented professional engravers and printers in the form of sumptuous color plates. A culture of patronage brought great success for many artists, authors and publishers, as popular and unusual plants—tulips, carnations, orchids and more—were captured on paper and vellum. Artists also illustrated native plants, creating sketchbooks and floras of local species.

As new plants from around the world were being collected and studied, the desire for knowledge about plants and their properties transformed medicinal botany into a new science focused on the sheer variety of the plants themselves. The study of plant form also took on new significance as a basis for new systems of classification. The "sexual system" developed by the Swedish botanist Carl von Linné required careful observation of the parts of flowers and was popularized by the illustrations of Georg Dionysius Ehret (1708–1770), one of the most impressive and prolific of eighteenth-century botanical artists. Based in London, Ehret produced work for many publications while also teaching and inspiring others. In Paris, Pierre-Joseph Redouté (1759–1840) continued the traditions of illustrious predecessors, such as Nicolas Robert (1614–1685), and achieved great success with his paintings of roses, lilies, and much else for the French elite. In the era of Ehret and Redouté, countless other brilliant and skillful artists were in great demand, especially in England, to support the burgeoning bibliographic production.

Coincident with the upsurge in demand that flowed from the interests of artists, gardeners, and scientists, the production of books and the dissemination of botanical art was revolutionized in the late 1700s and early 1800s with the invention of intaglio printing and the engraving and etching of copper plates. The resulting prints, often embellished by hand-coloring, adorned the books collected by wealthy patrons. In the mid-nineteenth century, lithography became a prominent publishing technique, followed by chromolithography. As a result, as the demand for art in scientific works continued to grow, publication became more available and affordable. Many leading botanists from this era were themselves accomplished artists— for example, William Jackson Hooker and his son Joseph Dalton Hooker, the first two Directors of the Royal Botanic Gardens, Kew—but they also employed a new generation of superb illustrators, including Walter Hood Fitch, who worked for both Hookers. And while many captivating plant portraits were specifically created for books and journals, many artists, both amateur and professional, created beautiful originals that were saved in sketch books or bound into unpublished manuscripts.

Through a combination of demand and technical advancement, botanical art reached its zenith during the nineteenth century, coincident with the life and work of Charles Darwin. Through his studies at Edinburgh and Cambridge, his voyage on the *Beagle*, and his extensive networks, Darwin gained an unusually broad knowledge of plants from many parts of the world. Later, living close to London and with strong connections at the Royal Botanic Gardens, Kew Darwin certainly had access to extraordinary works of art, and perhaps perused accurate depictions of the many plants that intrigued him. Yet Darwin included very few images in his books. Curious to see what botanical paintings would have been available during Darwin's time, and the great range of plants to which he potentially had access, we undertook the enjoyable task of gazing on his behalf upon some of the greatest botanical art of all time, especially through the unusually comprehensive collection in the library of Oak Spring Garden Foundation.

In the spirit of encouraging an appreciation of the art of Darwin's botanical science, it is appropriate to also encourage an appreciation of his subjects and of botanical art itself. Drawing upon the stunning botanical painting and print collection of Oak Spring Garden Foundation, we bring these complementary art forms together here for modern readers. We hope that doing so fosters an appreciation of the beauty of some of Darwin's favorite plants as well as the beauty of the scientific insights they yielded in the admiring hands of that inveterate "experimentiser."

Cyclamen europaeum (= *purpurascens*). Water and bodycolor on vellum by
Georg Dionysius Ehret, in *Flowers, Moths, Butterflies and Shells*.

Angraecum sesquipedale. Chromolithograph from Frederick Sander, *Reichenbachia.*

Angraecum
COMET ORCHID
———— ...... ————
ORCHIDACEAE—ORCHID FAMILY

ORCHIDS, FORMS OF FLOWERS, POLLINATION

Angraecum sesquipedale, often called Darwin's Comet Orchid, is one of
220 *Angraecum* species with star-shaped flowers bearing long nectar
spurs; many of the species are found in Madagascar. The flowers are
considered a spectacular example of coevolution, with pollinators
adapted to pollinate specific species. Darwin received a specimen of
A. *sesquipedale* in January 1862 from Staffordshire banker and orchid
enthusiast James Bateman. Marveling at the length of its nectar spur,
Darwin wrote to Joseph Hooker, "I have just received such a Box full
from Mr Bateman with the astounding Angraecum sesquipedalia
with a nectary a foot long.—Good Heavens what insect can suck it."[22]
Darwin did tests with long thin rods to remove pollen and hypothesized that this species must be pollinated by a moth with a proboscis
long enough to reach the nectar at the bottom of the long spur.

[**From: *The Various Contrivances by Which
Orchids are Fertilised by Insects* (2nd ed., 1877)**]

The nectar-secreting organs of the Orchideae present great diversities of
structure and position in the various genera; but are almost always situated
towards the base of the labellum. ...

The *Angraecum sesquipedale*, of which the large six-rayed flowers, like
stars formed of snow-white wax, have excited the admiration of travellers

in Madagascar, must not be passed over. A green, whip-like nectary of astonishing length hangs down beneath the labellum. In several flowers sent me by Mr. Bateman, I found the nectaries eleven and a half inches long, with only the lower inch and a half filled with nectar. What can be the use, it may be asked, of a nectary of such disproportionate length? We shall, I think, see that the fertilisation of the plant depends on this length, and on nectar being contained only within the lower and attenuated extremity. It is, however, surprising that any insect should be able to reach the nectar. Our English sphinxes have proboscides as long as their bodies; but in Madagascar there must be moths with proboscides capable of extension to a length of between ten and eleven inches! This belief of mine has been ridiculed by some entomologists, but we now know from Fritz Müller that there is a sphinx-moth in South Brazil which has a proboscis of nearly sufficient length, for when dried, it was between ten and eleven inches long. When not protruded, it is coiled up into a spiral of at least twenty windings.

I could not for some time understand how the pollinia of this Orchid were removed, or how the stigma was fertilised. I passed bristles and needles down the open entrance into the nectary and through the cleft in the rostellum with no result. It then occurred to me that, from the length of the nectary, the flower must be visited by large moths, with a proboscis thick at the base; and that to drain the last drop of nectar, even the largest moth would have to force its proboscis as far down as possible. Whether or not the moth first inserted its proboscis by the open entrance into the nectary, as is most probable from the shape of the flower, or through the cleft in the rostellum, it would ultimately be forced, in order to drain the nectary, to push its proboscis through the cleft, for this is the straightest course; and by slight pressure the whole foliaceous rostellum is depressed. The distance from the outside of the flower to the extremity of the nectary can be thus shortened by about a quarter of an inch. I therefore took a cylindrical rod one-tenth of an inch in diameter and pushed it down through the cleft in the rostellum. The margins readily separated and were pushed downwards together with the whole rostellum. When I slowly withdrew the cylinder, the rostellum rose from its elasticity, and the margins of the cleft were upturned so as to clasp the cylinder. Thus the viscid strips of membrane on each under side of the cleft rostellum came into contact with the cylinder, and firmly adhered to it; and the pollen-masses were withdrawn. By this means I succeeded every time in withdrawing the pollinia; and it cannot, I think, be doubted that a large moth would thus act; that is, it would drive its proboscis up to the very base through the cleft of the rostellum, so as to reach the extremity of the nectary; and then the pollinia attached to the base of its proboscis would be safely withdrawn.

I did not succeed in leaving the pollen-masses on the stigma so well as I did in withdrawing them. As the margins of the cleft rostellum must be upturned before the discs adhere to a cylindrical body, during its withdrawal, the pollen-masses become affixed some little way from its base. The two discs did not always adhere at exactly opposite points. Now, when a moth with the pollinia adhering to the base of its proboscis, inserts it for a second time into the nectary, and exerts all its force so as to push down the rostellum as far as possible, the pollen-masses will generally rest on and adhere to the narrow, ledge-like stigma which projects beneath the rostellum. By acting in this manner with the pollinia attached to a cylindrical object, the pollen-masses were twice torn off and left glued to the stigmatic surface.

If the *Angraecum* in its native forests secretes more nectar than did the vigorous plants sent me by Mr. Bateman, so that the nectary ever becomes filled, small moths might obtain their share, but they would not benefit the plant. The pollinia would not be withdrawn until some huge moth, with a wonderfully long proboscis, tried to drain the last drop. If such great moths were to become extinct in Madagascar, assuredly the *Angraecum* would become extinct. On the other hand, as the nectar, at least in the lower part of the nectary, is stored safe from the depredation of other insects, the extinction of the *Angraecum* would probably be a serious loss to these moths. We can thus understand how the astonishing length of the nectary had been acquired by successive modifications. As certain moths of Madagascar became larger through natural selection in relation to their general conditions of life, either in the larval or mature state, or as the proboscis alone was lengthened to obtain honey from the *Angraecum* and other deep tubular flowers, those individual plants of the *Angraecum* which had the longest nectaries (and the nectary varies much in length in some Orchids), and which, consequently, compelled the moths to insert their proboscides up to the very base, would be best fertilised. These plants would yield most seed, and the seedlings would generally inherit long nectaries; and so it would be in successive generations of the plant and of the moth. Thus it would appear that there has been a race in gaining length between the nectary of the *Angraecum* and the proboscis of certain moths; but the *Angraecum* has triumphed, for it flourishes and abounds in the forests of Madagascar, and still troubles each moth to insert its proboscis as deeply as possible in order to drain the last drop of nectar.

In response to Darwin's critics who doubted such a close relationship could evolve by natural selection, his friend and colleague Alfred Russel Wallace had undertaken a search for the theorized pollinator. Scouring the British Museum's insect collection for candidates, ruler in hand, Wallace had found sphinx moths of the genus *Macrosila* (now *Xanthopan*) from tropical Africa and South America with prosbosces as long as 9¼ inches. Writing in an 1867 review, Wallace declared, "A species having a proboscis two or three inches longer could reach the nectar in the largest flowers of *Angraecum sesquipedale*, whose nectaries vary in length from ten to fourteen inches. That such a moth exists in Madagascar may be safely predicted; and naturalists who visit that island should search for it with as much confidence as astronomers searched for the planet Neptune,—and they will be equally successful!"[23] The predicted moth, *Xanthopan morgani praedicta*, discovered in 1903, is a subspecies of one of the moths mentioned by Wallace in his article. It is fitting that Darwin's Comet Orchid should be pollinated by Wallace's Hawkmoth.

Arachis hypogaea. Hand-colored engraved plate drawn by M. M. Payerlein from
Christoph Jacob Trew, *Plantae Rariores Quas Maximam Partum.*

Arachis
PEANUT or GROUNDNUT

———— ...... ————

FABACEAE—PEA FAMILY

PLANT MOVEMENT

Arachis is a genus with nearly seventy species, growing in South America's dry tropical and subtropical grasslands. Only one is cultivated agriculturally, the common peanut *Arachis hypogaea*, a species of hybrid origin that is thought to have arisen several millennia ago in the Andean region. Peanuts are now grown commercially in warm temperate areas around the world, a major human food source but also animal fodder and ground cover.

The peanut genus is derived from the Greek word for vetch, *arakos*, also a legume, while the species epithet *hypogaea*, "beneath the earth," refers to the curious habit of geocarpy, fruit development underground—perhaps originally a nifty adaptation for both safeguarding seeds from predation and automatically planting them. The yellow flowers bear what appears to be an ordinary stalk, or peduncle, but is actually a long calyx tube (hypanthium), at the base of which is the ovary. Peanut flowers mainly self-pollinate, a common feature in many legumes, and the pollen tubes successfully make their way down to the basal ovary. They then produce an elongated "peg" (the "gynophore" in Darwin's terms), a stalk-like structure that extends from beneath the ovary, bearing the developing seed pod down into the soil where the fruits develop. With the help of William Thiselton-Dyer, assistant director of the Royal Botanic Gardens, Kew, Darwin procured potted peanut plants for study. In the course of investigating what he called *geotropism* and *apogeotropism*—plant movement toward or away from the earth, respectively, in response to gravity—he traced their movement on a vertically held pane of glass, finding that the growing pegs circumnutate, slowly rotating in an elliptical path as they grow downward.

[**From: *The Power of Movement in Plants* (1880)**]

Arachis hypogaea—The shape of a leaf, with its two pairs of leaflets, is shown at A; and a leaf asleep, traced from a photograph (made by the aid of aluminium light), is given at B. The two terminal leaflets twist round at night until their blades stand vertically, and approach each other until they meet, at the same time moving a little upwards and backwards. The two lateral leaflets meet each other in this same manner, but move to a greater extent forwards, that is, in a contrary direction to the two terminal leaflets, which they partially embrace. Thus all four leaflets form together a single packet, with their edges directed to the zenith, and with their lower surfaces turned outwards. ... The petioles are inclined upwards during the day, but sink at night, so as to stand at about right angles with the stem. The amount of sinking was measured only on one occasion and found to be 39°. A petiole was secured to a stick at the base of the two terminal leaflets,

Arachis hypogaea: A, leaf during the day, seen from vertically above; B, leaf asleep, seen laterally; copied from a photograph. Figures much reduced.

and the circumnutating movement of one of these leaflets was traced from 6.40 A.M. to 10.40 P.M., the plant being illuminated from above. ... During the 16 h. the leaflet moved thrice up and thrice down, and as the ascending and descending lines did not coincide, three ellipses were formed. ...

The flowers, which bury themselves, rise from stiff branches a few inches above the ground, and stand upright. After they have fallen off, the gynophore, that is the part which supports the ovarium, grows to a great length, even to 3 or 4 inches, and bends perpendicularly downwards. It resembles closely a peduncle, but has a smooth and pointed apex, which contains the ovules, and is at first not in the least enlarged. The apex after reaching the ground penetrates it, in one case observed by us to a depth of 1 inch, and in another to 0.7 inch. It there becomes developed into a large pod. Flowers which are seated too high on the plant for the gynophore to reach the ground are said never to produce pods.

The movement of a young gynophore, rather under an inch in length and vertically dependent, was traced during 46 h. by means of a glass filament (with sights) fixed transversely a little above the apex. It plainly circumnutated ... whilst increasing in length and growing downwards. It was then raised up, so as to be extended almost horizontally, and the terminal part curved itself downwards, following a nearly straight course during 12 h., but with one attempt to circumnutate, as shown. ... After 24 h. it had become nearly vertical. Whether the exciting cause of the downward movement is geotropism or apheliotropism was not ascertained; but probably it is not apheliotropism, as all the gynophores grew straight down towards the ground, whilst the light in the hot-house entered from one side as well as from above. Another and older gynophore, the apex of which had nearly reached the ground, was observed during 3 days in the same manner as the first-mentioned short one; and it was found to be always circumnutating. During the first 34 h. it described a figure which represented four ellipses.

Lastly, a long gynophore, the apex of which had buried itself to the depth of about half an inch, was pulled up and extended horizontally: it quickly began to curve downwards in a zigzag line; but on the following day

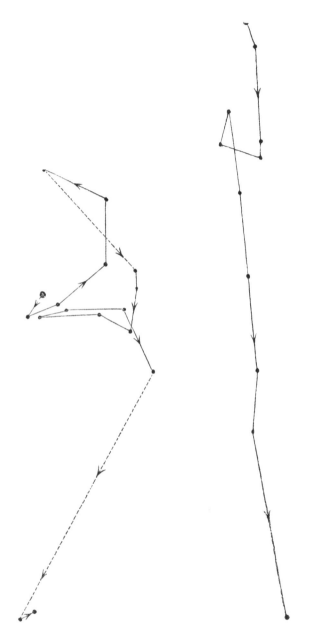

Arachis hypogaea: circumnutation of vertically dependent young gynophore, traced on a vertical glass from 10 A.M. July 31st to 8 A.M. Aug. 2nd.

Arachis hypogaea: downward movement of same young gynophore, after being extended horizontally; traced on a vertical glass from 8.30 A.M. to 8.30 P.M. Aug. 2nd.

the terminal bleached portion was a little shrivelled. As the gynophores are rigid and arise from stiff branches, and as they terminate in sharp smooth points, it is probable that they could penetrate the ground by the mere force of growth. But this action must be aided by the circumnutating movement, for fine sand, kept moist, was pressed close round the apex of a gynophore which had reached the ground, and after a few hours it was surrounded by a narrow open crack. After three weeks this gynophore was uncovered, and the apex was found at a depth of rather above half an inch developed into a small, white, oval pod.

Darwin was also interested in nyctinasty, the nocturnal "sleep" movement of peanut leaves as they fold their leaves at night. To test the hypothesis that nocturnal leaf-folding was an adaptation to reduce frost damage through radiative heat loss (by avoiding direct exposure of the leaf surface to the night sky), Darwin and his son Francis set out potted peanut plants in the open lawn behind their house on freezing late winter nights, preventing some of the leaves from folding by various means, while letting others move freely. By and large, all the leaves kept from folding were frost-killed or injured, while only about half of the "control" leaves suffered in this way.

Darwin excitedly reported the results to Joseph Hooker at Kew, saying, "I think we have *proved* that the sleep of plants is to lessen injury to leaves from radiation—This has interested me much and has cost us great labour, as it has been a problem since the time of Linnaeus."[24] The reference here was to a doctoral dissertation, *Somnus plantarum* ("The Sleep of Plants"), supervised by Linnaeus and published in 1755. Linnaeus thought plants did indeed simply sleep, but Darwin and his son now had evidence of an adaptive explanation for nocturnal leaf movement. Their successes came at a cost, however, as Darwin lamented in the same letter, "we have killed or badly injured a multitude of plants," urging Hooker to send more as soon as possible. "Unfortunately, there is no time to lose, as there may be few more frosts."

Clematis Americana
siliquosa tetraphyllos.

Clematis d'Amerique a quatre
feuilles, portant des gousses.

N. Robert sculp.

Bignonia capreolata. Engraved and etched plate by Nicolas Robert, from Denis
Dodart, *Mémoires pour Servir à l'Histoire des Plantes*.

Bignonia

CROSSVINE, TRUMPET
CREEPER, and RELATIVES

——————— ...... ———————

BIGNONIACEAE—BIGNONIA FAMILY

CLIMBING PLANTS

Bignoniaceae is a large and mainly tropical family of trees, shrubs, and lianas (woody vines). The climbers of the family most interested Darwin, and in distinguishing between types of climbing, those endowed with what he called "irritable organs" (modified leaves, branches, or flower peduncles sensitive to touch) especially impressed him. The New World type-genus *Bignonia*, the name of which honors the Frenchman Abbé Jean-Paul Bignon (1662–1743), checked all these boxes for him, sporting "irritable" internodes, petioles, and tendrils. Many members of this genus of vigorous woody vines are prized horticulturally for their large trumpet-shaped flowers. He studied nearly a dozen species,* taking what he could get from his contacts at Kew Gardens and Veitch's Royal Exotic Nursery in London.

Bignonia capreolata, known as crossvine, is native to the southern and central United States and is one of the most remarkable species Darwin grew. This species was well known in English horticultural circles as one of the earliest botanical imports from the American

---

* Most of the Bignoniaceae that Darwin studied were placed in the genus *Bignonia* at the time, but there have been many taxonomic changes since then. *Bignonia unguis* and *B. tweedyana* = *Dolichandra unguis-cati*; *Bignonia venusta* = *Pyrostegia venusta*; *Bignonia littoralis* = *Fridericia mollissima*; *Bignonia speciosa* = *Bignonia callistegioides*; *Bignonia picta* = *Bignonia aequinoctialis*.

colonies. Darwin noticed that they tend to grow away from light, which seemed counterintuitive—shouldn't vines always reach for the sun? In one experiment, he placed a potted crossvine with six tendrils into a box open on one side and positioned the open side obliquely to a light source. Checking on the plant two days later, he found that the tendrils were all reaching toward the darkest corner of the box. "Six wind-vanes could not have more truly shown the direction of the wind," he declared, "than did these branched tendrils the course of the stream of light which entered the box."[25] This vine's shade-seeking behavior—*apheliotropism*, as the tendency to move away from light is called—is thought to help young earthbound vines on the forest floor locate a tree to climb, a surer path to getting upwards to the needed light rather than responding to light at ground level.

Through trial-and-error, Darwin also found that this species seems to prefer climbing a rough woolly-textured surface. They refused, so to speak, smooth sticks but readily climbed when Darwin wrapped the sticks in flax, moss, or wool. Darwin surmised that this preference was related to the plant's native environment and put the question to his botanist friend Asa Gray at Harvard in the United States. "Have you travelled south, and can you tell me, whether the trees, which *Bignonia capreolata* climbs, are covered with moss, or filamentous lichen or Tillandsia; I ask because its tendrils abhor a simple stick, do not much relish rough bark, but delight in wool or moss," he wrote, anthropomorphizing to evoke the "animal" in the plant, as he often did. Gray confirmed that the vines did indeed climb trees "well furnished with Lichens and Mosses" in the wet, shady forests of the southern United States.[26] Darwin found that several other *Bignonia* species, as well as some species in the allied genera *Tecoma* and *Eccremocarpus*, also grow away from sunlight to attach themselves for support to walls or trees.

Note how, near the end of Darwin's account of his experiments with *B. capreolata*, excerpted here, he steps back to marvel at the significance of his findings: insofar as tendrils are highly modified leaves, adapted to seek light, it's amazing that here they should evolve into a structure that moves *away* from light, seeking, root-like, nooks and crannies to grab hold of.

[**From:** ***The Movements and Habits of Climbing Plants*** **(2nd ed., 1875)**]

Bignonia unguis.— ... The stem twines imperfectly round a vertical stick, sometimes reversing its direction in the same manner as described in so many leaf-climbers; and this plant though possessing tendrils, climbs to a certain extent like a leaf-climber. Each leaf consists of a petiole bearing a pair of leaflets, and terminates in a tendril, which is formed by the modification of three leaflets, and closely resembles that above figure. ... It is curiously like the leg and foot of a small bird, with the hind toe cut off. The straight leg or tarsus is longer than the three toes, which are of equal length, and diverging, lie in the same plane. The toes terminate in

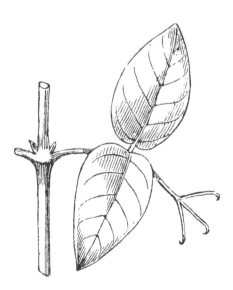

Bignonia Unnamed species from Kew

sharp, hard claws, much curved downwards, like those on a bird's foot. The petiole of the leaf is sensitive to contact; even a small loop of thread suspended for two days caused it to bend upwards; but the sub-petioles of the two lateral leaflets are not sensitive. The whole tendril, namely, the tarsus and the three toes, are likewise sensitive to contact, especially on their under surfaces. When a shoot grows in the midst of thin branches, the tendrils are soon brought by the revolving movement of the internodes into contact with them; and then one toe of the tendril or more, commonly all three, bend, and after several hours seize fast hold of the twigs, like a bird when perched. If the tarsus of the tendril comes into contact with a twig, it goes on slowly bending, until the whole foot is carried quite round, and the toes pass on each side of the tarsus and seize it. In like manner, if the petiole comes into contact with a twig, it bends round, carrying the tendril, which then seizes its own petiole or that of the opposite leaf. The petioles move spontaneously, and thus, when a shoot attempts to twine round an upright stick, those on both sides after a time come into contact with it and are excited to bend. Ultimately the two petioles clasp the stick in opposite directions, and the foot-like tendrils, seizing on each other or on their own petioles, fasten the stem to the support with surprising security. ... This plant is one of the most efficient climbers which I have observed; and it probably could ascend a polished stem incessantly tossed by heavy storms. ...

Bignonia capreolata.—We now come to a species having tendrils of a different type; but first for the internodes. A young shoot made three large revolutions, following the sun, at an average rate of 2 hrs. 23 m. The stem is thin and flexible, and I have seen one make four regular spiral turns round a thin upright stick, ascending of course from right to left, and therefore in a reversed direction compared with the before described species. Afterwards, from the interference of the tendrils, it ascended either straight up the stick or in an irregular spire. The tendrils are in some respects highly remarkable. In a young plant they were about 2½ inches in length and much branched, the five chief branches apparently representing two pairs of leaflets and a terminal one ... with the points blunt yet distinctly hooked. ...

Whilst the tendrils are revolving more or less regularly, another remarkable movement takes place, namely, a slow inclination from the light towards the darkest side of the house. I repeatedly changed the position of my plants, and some little time after the revolving movement had ceased, the successively formed tendrils always ended by pointing to the darkest side. When I placed a thick post near a tendril, between it and the light, the tendril pointed in that direction. In two instances a pair of leaves stood so that one of the two tendrils was directed towards the light and the other to the darkest side of the house; the latter did not move, but the opposite one bent itself first upwards and then right over its fellow, so that the two became parallel, one above the other, both pointing to the dark: I then turned the plant half round; and the tendril which had turned over recovered its original position, and the opposite one which had not before moved, now turned over to the dark side. Lastly, on another plant, three pairs of tendrils were produced at the same time by three shoots, and all happened to be differently directed: I placed the pot in a box open only on one side, and obliquely facing the light; in two days all six tendrils pointed with unerring truth to the darkest corner of the box, though to do this each had to bend in a different manner. Six wind-vanes could not have more truly shown the direction of the wind, than did these branched tendrils the course of the stream of light which entered the box. ...

When a tendril has not succeeded in clasping a support, either through its own revolving movement or that of the shoot, or by turning towards any object which intercepts the light, it bends vertically downwards and then towards its own stem, which it seizes together with the supporting stick, if there be one. A little aid is thus given in keeping the stem secure. If the tendril seizes nothing, it does not contract spirally, but soon withers away and drops off. If it seizes an object, all the branches contract spirally.

I have stated that after a tendril has come into contact with a stick, it bends round it in about half an hour; but I repeatedly observed, as in the case of *B. speciosa* and its allies, that it often again loosed the stick; sometimes seizing and loosing the same stick three or four times. Knowing that the tendrils avoided the light, I gave them a glass tube blackened within, and a well-blackened zinc plate: the branches curled round the tube and abruptly bent themselves round the edges of the zinc plate; but they soon recoiled from these objects with what I can only call disgust and straightened themselves. I then placed a post with extremely rugged bark close to a pair of tendrils; twice they touched it for an hour or two, and twice they withdrew; at last one of the hooked extremities curled round and firmly seized an excessively minute projecting point of bark, and then the other branches spread themselves out, following with accuracy every inequality of the surface. I afterwards placed near the plant a post without bark but much fissured, and the points of the tendrils crawled into all the crevices in a beautiful manner. To my surprise, I observed that the tips of the immature tendrils, with the branches not yet fully separated, likewise crawled just like roots into the minutest crevices. In two or three days after the tips had thus crawled into the crevices, or after their hooked ends had seized minute points, the final process, now to be described, commenced.

This process I discovered by having accidentally left a piece of wool near a tendril; and this led me to bind a quantity of flax, moss, and wool loosely round sticks, and to place them near tendrils. The wool must not be dyed, for these tendrils are excessively sensitive to some poisons. The hooked points soon caught hold of the fibres, even loosely floating fibres, and now there was no recoiling; on the contrary, the excitement caused the hooks to penetrate the fibrous mass and to curl inwards, so that each hook caught firmly one or two fibres, or a small bundle of them. The tips and the inner surfaces of the hooks now began to swell, and in two or three days were visibly enlarged. After a few more days the hooks were converted into whitish, irregular balls, rather above the $\frac{1}{20}$th of an inch (1.27 mm.) in diameter, formed of coarse cellular tissue, which sometimes wholly enveloped and concealed the hooks themselves. The surfaces of these balls secrete some viscid resinous matter, to which the fibres of the flax, &c., adhere. When a fibre has become fastened to the surface, the cellular tissue does not grow directly beneath it, but continues to grow closely on each side; so that when several adjoining fibres, though excessively thin, were caught, so many crests of cellular matter, each not as thick as a human hair, grew up between them, and these, arching over on both sides, adhered firmly together. As the whole surface of the ball continues to grow, fresh fibres adhere and are afterwards enveloped; so that I have seen a little ball with between fifty and sixty fibres of flax crossing it at various angles and all embedded more or less deeply. Every gradation in the process could be followed—some fibres merely sticking to the surface, others lying in more or less deep furrows, or deeply embedded, or passing through the very centre of the cellular ball. ...

From the facts now given, we may infer that though the tendrils of this *Bignonia* can occasionally adhere to smooth cylindrical sticks and often to rugged bark, yet that they are specially adapted to climb trees clothed with lichens, mosses, or other such productions; and I hear from Professor Asa Gray that the *Polypodium incanum* abounds on the forest-trees in the districts of North America where this species of *Bignonia* grows. Finally, I may remark how singular a fact it is that a leaf should be metamorphosed into a branched organ which turns from the light, and which can by its extremities either crawl like roots into crevices, or seize hold of minute projecting points, these extremities afterwards forming cellular outgrowths which secrete an adhesive cement, and then envelop by their continued growth the finest fibres.

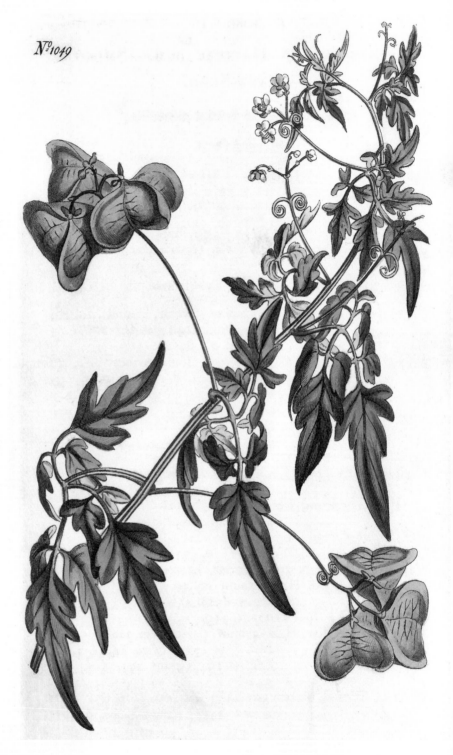

Cardiospermum halicacabum. Hand-colored engraving, drawn by Sydenham Edwards, from *The Botanical Magazine* 26: 1049.

```
┌─────────────────────────────────────────────────────┐
│                                                       │
│                  Cardiospermum                        │
│               BALLOON VINE or                         │
│               LOVE-IN-A-PUFF                          │
│            ───────  ......  ───────                   │
│                                                       │
│          SAPINDACEAE—SOAPBERRY FAMILY                 │
│                                                       │
└─────────────────────────────────────────────────────┘
```

CLIMBING PLANTS

A genus of about fifteen species, *Cardiospermum* is native to the American tropics, with a few species with pantropical distribution and several prized world-wide as ornamentals. All of the species are herbaceous tendrilled vines with flowers produced in compact axillary panicles, each bearing a pair of tendrils at the base. The distinctive feature of the genus is the fruit, an inflated three-lobed pod that inspired the common name "balloon vine," but equally appealing are the black seeds hidden within, each bearing a distinctive white heart-shaped mark—hence "Love-in-a-puff" (and the genus name, *Cardiospermum*, derived from Latin for "heart-seed").

Cardiospermum is in the soapberry family, and the seeds are certainly "loved" by beautiful soapberry bugs, a group of some sixty-five species making up a subfamily of the uninspiringly named scentless plant bug family (Rhopalidae). Darwin would have been intrigued to learn that certain soapberry bugs have become models for the study of rapid evolutionary adaptation, thanks to humans planting *Cardiospermum*, soapberry (*Sapindus saponaria*), and their relatives far and wide—within mere decades of introducing new hosts, native soapberry bugs evolve different beak lengths to feed on the new and different-sized fruits, growth rates, and even hostplant preferences.*[27]

Darwin's interest, however, was in the curious tendrils of these plants. At the height of his fascination with tendrils, Darwin wrote

Joseph Hooker at the Royal Botanic Gardens, Kew asking if his friend could "by an extraordinary good chance give me *now* a plant of *Cardiospermum halicacabum* (or any other species if such has tendrils)."[28] Hooker was always a great help, and Darwin got his plant. Over time he worked out that, like in *Vitis* (see p. 333), *Cardiospermum* tendrils are modified flower peduncles—a diagnosis clinched by finding rare cases where the tendrils themselves actually yielded flowers, a curious sport. He also speculated that, besides climbing, the tendrils have a secondary role in helping to secure the large pods so that they don't get blown about and damaged by the wind.

[**From: *The Movements and Habits of Climbing Plants* (2nd ed., 1875)**]

Cardiospermum halicacabum.—In this family, as in Vitaceae, the tendrils are modified flower-peduncles. In the present plant the two lateral branches of the main flower-peduncle have been converted into a pair of tendrils, corresponding with the single "flower-tendril" of the common vine. The main peduncle is thin, stiff, and from 3 to 4½ inches in length. Near the summit, above two little bracts, it divides into three branches. The middle one divides and re-divides and bears the flowers; ultimately it grows half as long again as the two other modified branches. These latter are the tendrils; they are at first thicker and longer than the middle branch, but never become more than an inch in length. They taper to a point and are flattened, with the lower clasping surface destitute of hairs. At first they project straight up; but soon diverging, spontaneously curl downwards

---

* (previous) The best-studied example is the red-shouldered bug (*Jadera haematoloma*), the native host plant of which is the balloon vine (*Cardiospermum corindum*). Both plant and bug are denizens of Florida hammocks, but the insect has rapidly evolved to take advantage of a new non-native host, the golden rain tree (*Koelreuteria elegans*), introduced from Asia and now widely found in Florida.

Cardiospermum halicacabum. Upper part of
the flower-peduncle with its two tendrils.

so as to become symmetrically and elegantly hooked, as represented in
the diagram. They are now, whilst the flower-buds are still small, ready
for action.

The two or three upper internodes, whilst young, steadily revolve; those
on one plant made two circles against the course of the sun in 3 hrs. 12 m.;
in a second plant the same course was followed, and the two circles were
completed in 3 hrs. 41 m.; in a third plant, the internodes followed the sun
and made two circles in 3 hrs. 47 m. The average rate of these six revolutions was 1 hr. 46 m. The stem shows no tendency to twine spirally round a
support; but the allied tendril-bearing genus *Paullinia* is said to be a twiner.
The flower-peduncles, which stand up above the end of the shoot, are
carried round and round by the revolving movement of the internodes;
and when the stem is securely tied, the long and thin flower-peduncles
themselves are seen to be in continued and sometimes rapid movement
from side to side. They sweep a wide space, but only occasionally revolve
in a regular elliptical course. By the combined movements of the internodes and peduncles, one of the two short, hooked tendrils, sooner or

later catches hold of some twig or branch, and then it curls round and securely grasps it. These tendrils are, however, but slightly sensitive; for by rubbing their under surface only a slight movement is slowly produced. I hooked a tendril on to a twig; and in 1 hr. 45 m. it was curved considerably inwards; in 2 hrs. 30 m. it formed a ring; and in from 5 to 6 hours from being first hooked, it closely grasped the stick. A second tendril acted at nearly the same rate; but I observed one that took 24 hours before it curled twice round a thin twig. Tendrils which have caught nothing, spontaneously curl up to a close helix after the interval of several days. Those which have curled round some object, soon become a little thicker and tougher. The long and thin main peduncle, though spontaneously moving, is not sensitive and never clasps a support. Nor does it ever contract spirally, although a contraction of this kind apparently would have been of service to the plant in climbing. Nevertheless it climbs pretty well without this aid. The seed-capsules, though light, are of enormous size (hence its English name of balloon-vine), and as two or three are carried on the same peduncle, the tendrils rising close to them may be of service in preventing their being dashed to pieces by the wind. In the hothouse, the tendrils served simply for climbing.

The position of the tendrils alone suffices to show their homological nature. In two instances, one of two tendrils produced a flower at its tip; this, however, did not prevent its acting properly and curling round a twig. In a third case, both lateral branches, which ought to have been modified into tendrils, produced flowers like the central branch and had quite lost their tendril-structure.

Catasetum saccatum. Chromolithograph drawn by A. Goessens,
from Jean Jules Linden, *Lindenia*, *Iconography of Orchids*.

CATASETUM ORCHIDS

------ ...... ------

ORCHIDACEAE—ORCHID FAMILY

ORCHIDS, FORMS OF FLOWERS, POLLINATION

Catasetum orchids are remarkably complex and elaborate, with 130 species found in Neotropical regions. It is one of the few orchid groups that bear separate male and female flowers, with a dramatic sexual dimorphism that fooled botanists for decades into thinking they belonged to different genera. Even more exciting for Darwin, the large male flowers forcibly discharge their sticky pollen packets (pollinia) in a form of "ballistic" pollen transfer that had never before been seen in any orchid.[29]

When Darwin received a *Catasetum* plant in bloom from James Veitch's Royal Exotic Nursery in 1861, he declared to Joseph Hooker that it was the most wonderful orchid he had ever seen. He was in good company—nearly forty years earlier William Jackson Hooker, Joseph's father and director of Kew before him, had declared of *Catasetum tridentatum*, "I know of no individual of [the orchid] family which has flowers so splendid and so curious."[30] *Catasetum* inflamed Darwin's growing orchidelirium, and he wrote to Hooker, "I never was more interested in any subject in my life, than in this of Orchids."[31] The "special contrivance" of the amazing pollinia ejection led him to dub *Catasetum* "the most remarkable of all orchids,"[32] and the jokester couldn't resist firing salvos of *Catasetum* pollen at unsuspecting guests visiting his greenhouse.

Catasetum was considered a genus wholly distinct from two other unusual genera in the family, *Monachanthus* and *Myanthus*—orchids that, incredibly, were found growing on the same individual plant. Naturalist and explorer Robert Hermann Schomburgk first presented specimens of this Franken-orchid to the Linnean Society in 1836. He soon collected another along the banks of the Essequibo River in South America, this one bearing flowers of *Catasetum tridentatum* and *Monachanthus viridis*. The keen observer noted that none of the hundreds of *C. tridentatum* flowers bore seeds, while he was "astonished" by the large *M. viridis* fruits. That proved to be the clue Darwin used to solve the mystery; he determined that these were, in fact, floral morphs of one and the same species. *Catasetum tridentatum* turned out to be male, *Monachanthus viridis* female (complete with vestigial pollinia!), and *Myanthus barbatus* a hermaphrodite with male and female parts.[33]

Darwin discussed these and related species at length in *The Various Contrivances by Which Orchids are Fertilised by Insects*. Hermann Crüger, a German pharmacist and botanist in Trinidad who had initially insisted the plants were distinctive genera, later confirmed Darwin's results and sent him specimens of the bees that pollinate them. But the perhaps slightly embarrassed Crüger could not resist taking a jab at Darwin too. "Whoever has read Darwin's remarkable work on the fecundation of orchids," he wrote in a letter to the Linnean Society of London,[34] "must have regretted that the chapters of tropical and other foreign orchids leave a certain amount of uncertainty on the mind of the reader until the observations and suppositions shall have been endorsed by actual facts observed in the native countries of these plants"—conveniently overlooking the fact that Darwin observed more in the preserved plants than those who had seen them live in the field. Darwin's conclusions were revised several years later, as often happens in science, but his observations corrected much of what other botanists had presented previously.

George Sowerby illustrated the wonderful flowers of *Catasetum* and other orchids for Darwin's book, aiding in his observations and descriptions. Sowerby spent ten days with Darwin, painstakingly illustrating flowers sent from Kew and recreating illustrations by botanical artist Franz Bauer. Darwin sighed in one letter that he was "half dead working with Mr. Sowerby at the orchid drawings,"[35] but the results were well worth the time.

[From: *The Various Contrivances by Which Orchids are Fertilised by Insects* (2nd ed., 1877)]

Catasetum tridentatum. The general appearance of this species, which is very different from that of *C. saccatum, callosum and tabulare*, is represented [here] with a sepal on each side cut off.

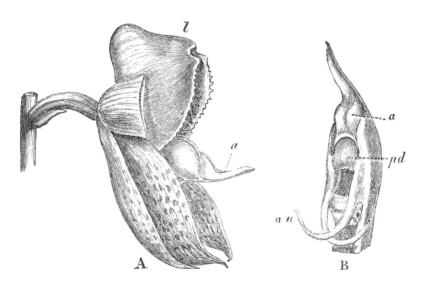

Catasetum tridentatum a. anther. *pd.* pedicel of pollinium. *an.* antennae. *l.* labellum A. Side view of flower in its natural position, with two of the sepals cut off. B. Front view of column, in position reverse of fig. A.

The flower stands with the labellum uppermost, that is, in a reversed position compared with most Orchids. The labellum is helmet-shaped, its distal portion being reduced to three small points. It cannot hold nectar from its position; but the walls are thick, and have, as in the other species, a pleasant nutritious taste. The stigmatic chamber, though functionless as a stigma, is of large size. The summit of the column, and the spike-like anther, are not so much elongated as in *C. saccatum*. In other respects, there is no important difference. The antennae are of greater length; their tips for about one-twentieth of their length are roughened by cells produced into papillae. ...

I need not further describe the present species, except as to the position of the antennae. They occupied exactly the same position in all the many flowers which were examined. Both lie curled within the helmet-like labellum; the left-hand one stands higher up, with its inwardly bowed extremity in the middle; the right-hand antenna lies lower down and crosses the whole base of the labellum, with the tip just projecting beyond the left margin of the base of the column. Both are sensitive, but apparently the one which is coiled within the middle of the labellum is the more sensitive of the two. From the position of the petals and sepals, an insect visiting the flower would almost certainly alight on the crest of the labellum; and it could hardly gnaw any part of the great cavity without touching one of the two antennae, for the left-hand one guards the upper part, and the right-hand one the lower part. When either of these is touched, the pollinium is ejected and the disc will strike the head or thorax of the insect.

The position of the antennae in this *Catasetum* may be compared with that of a man with his left arm raised and bent so that his hand stands in front of his chest, and with his right arm crossing his body lower down so that the fingers project just beyond his left side. ...

Catasetum tridentatum is interesting under another point of view. Botanists were astonished when Sir R. Schomburgk stated that he had seen three forms, believed to constitute three distinct genera, namely, *Catasetum tridentatum, Monachanthus viridis,* and *Myanthus barbatus,* all growing on the same plant. Lindley remarked that "such cases shake to the foundation all our ideas of the stability of genera and species." Sir R. Schomburgk affirms that he has seen hundreds of plants of *C. tridentatum* in Essequibo without ever finding one specimen with seeds, whereas he was surprised at the gigantic seed-vessels of the *Monachanthus*; and he correctly remarks that "here we have traces of sexual difference in Orchideous flowers." Dr. Crüger also informs me that in Trinidad he never saw capsules naturally produced by the flowers of this *Catasetum*; nor when they were fertilised by him with their own pollen, as was done repeatedly. On the other hand, when he fertilised the flowers of the *Monachanthus viridis* with pollen from the *Catasetum*, the operation never failed. The *Monachanthus* also commonly produces fruit in a state of nature. ...

With respect to *Monachanthus viridis*, and *Myanthus barbatus*, the President of the Linnean Society has kindly permitted me to examine the spike bearing these two so-called genera, preserved in spirits, which was sent home by Sir R. Schomburgk. The flower of the *Monachanthus* (A) resembles pretty closely in external appearance that of *Catasetum tridentatum*. ... The labellum, which holds the same relative position to the other parts, is not nearly so deep, especially on the sides, and its edge is crenated. The other petals and sepals are all reflexed and are not so much spotted as in the *Catasetum*. The bract at the base of the ovarium is much larger. The whole column, especially the filament and the spike-like anther, are much shorter; and the rostellum is much less protuberant. The antennae are entirely absent, and the pollen-masses are rudimentary. These are interesting facts, from corroborating the view taken of the function of the antennae; for as there are no pollinia to eject, an organ adapted to convey the stimulus from the touch of an insect to the rostellum would be useless. I could find no trace of a viscid disc or pedicel, and no doubt they had been lost; for Dr. Crüger says that "the anther of the female flower drops off immediately after the opening of the same, i. e. before the flower has reached perfection as regards colour, size, and smell. The disc does not cohere, or very slightly, to the pollen-masses, but drops off about the same time, with the anther;" leaving behind them the rudimentary pollen-masses.

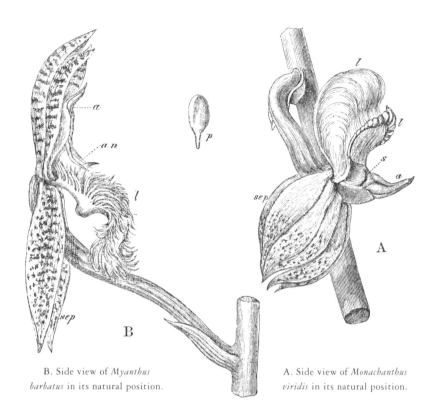

B. Side view of *Myanthus barbatus* in its natural position.

A. Side view of *Monachanthus viridis* in its natural position.

a. anther. *an.* antennae. *l.* labellum. *p.* pollen-mass, rudimentary.
s. stigma cleft. *sep.* two lower sepals.

From these facts alone, it is almost certain that *Monachanthus* is a female plant; and as already stated, Sir E. Schomburgk and Dr. Crüger have both seen it seeding abundantly. Altogether the flower differs in a most remarkable manner that that of the male *Catasetum tridentatum*, and it is no wonder that the two plants were formerly ranked as distinct genera. ...

Thus every detail of structure which characterises the male pollen-masses is represented in the female plant in a useless condition. Such cases are familiar to every naturalist but can never be observed without renewed interest. At a period not far distant, naturalists will hear with surprise, perhaps with derision, that grave and learned men formerly maintained that such useless organs were not remnants retained by inheritance but were specially created and arranged in their proper places like dishes on a table (this is the simile of a distinguished botanist) by an Omnipotent hand "to complete the scheme of nature."...

The genus *Catasetum* is interesting to an unusual degree in several respects. The separation of the sexes is unknown amongst other Orchids, except perhaps in the allied genus *Cycnoches*. In *Catasetum* we have three sexual forms, generally borne on separate plants, but sometimes mingled together on the same plant; and these three forms are wonderfully different from one another, much more different than, for instance, a peacock is from a peahen. ...

This genus is still more interesting in its manner of fertilisation. We see a flower patiently waiting with its antennae stretched forth in a well-adapted position, ready to give notice whenever an insect puts its head into the cavity of the labellum. The female *Monachanthus*, not having true pollinia to eject, is destitute of antennae. In the male and hermaphrodite forms, namely *Catasetum tridentatum* and *Myanthus barbatus*, the pollinia lie doubled up, like a spring, ready to be instantly shot forth when the antennae are touched. The disc end is always projected foremost and is coated with viscid matter which quickly sets hard and affixes the hinged pedicel firmly to the insect's body. The insect flies from flower to flower, till at last it visits a female plant: it then inserts one of the pollen-masses into the stigmatic cavity. As soon as the insect flies away the elastic caudicle, made weak enough to yield to the viscidity of the stigmatic surface, breaks, and leaves behind a pollen-mass; then the pollen-tubes slowly protrude, penetrate the stigmatic canal, and the act of fertilisation is completed. Who would have been bold enough to have surmised that the propagation of a species depended on so complex, so apparently artificial, and yet so admirable an arrangement?

Red creeping Climber.

Clematis repens. Water and bodycolor on vellum by Dame Ann Hamilton, *Drawings of Plants*.

CLIMBING PLANTS

Clematis is a genus of climbers admired since antiquity, the very name derived from one of the Ancient Greek words for vine: κλῆμα, *klema*. There are some 300 *Clematis* species worldwide, and a multitude of cultivars grown for their showy flowers. Darwin's interest had more to do with their leaves, however—specifically, the petioles, the thin stalks that attach a leaf to a stem. Writing to his friend Joseph Hooker in early 1864 to wheedle more specimens from Kew for study, Darwin explained what he had found so far: "when by growth the [petioles] of leaves are brought into contact with any object they bend and catch hold. The slightest stimulus suffices, even a bit of cotton thread a few inches long; but the stimulus must be applied during 6 or 12 hours, and when the [petioles] once bend, though touching object be removed, they never get straight again."[36] Hooker dutifully sent more species. Darwin was able to show that most *Clematis* vines are premier "leaf climbers," relying on their petioles to twist around and clasp onto whatever they can for support to reach heights. The petiole's touch-sensitivity could even betray the identity of rather un-*Clematis*-like members of the genus; in another letter to Hooker later that year Darwin remarked how, when he received as a gift a flowerless specimen of *Clematis glandulosa* (now *smilacifolia*), his gardener was initially dubious that it was a Clematis. "So," Darwin said, "I put a little twig to the [petiole] and the next day my gardener said 'you see it is a Clem[atis] for it feels.'... That's the way we make out plants at Down,"[37] he quipped.

Considered by Darwin to be intermediate between twiners and tendril bearers, leaf climbers are represented in various families, with *Clematis* species being especially responsive. He enjoyed observing their growth patterns, even as he himself suffered from various ailments. Working with eight species, he tested the reactions of the petioles under various circumstances, remarking on the "nervous system" that responded to touch by bending, holding, and expanding, growing thicker and stronger. Noting the minutes and hours it took the petioles to twine, he also measured and weighed bits of string that they clasped on to, emphasizing the sensitive nature of the petioles.

[**From: *The Movements and Habits of Climbing Plants***
(2nd ed., 1875)]

In all the species observed by me, with one exception, the young internodes revolve more or less regularly, in some cases as regularly as those of a twining plant. They revolve at various rates, in most cases rather rapidly. Some few can ascend by spirally twining round a support. Differently from most twiners, there is a strong tendency in the same shoot to revolve first in one and then in an opposite direction. The object gained by the revolving movement is to bring the petioles into contact with surrounding objects; and without this aid the plant would be much less successful in climbing. With rare exceptions, the petioles are sensitive only whilst young. They are sensitive on all sides, but in different degrees in different plants; and in some species of *Clematis* the several parts of the same petiole differ much in sensitiveness. ... The petioles are sensitive to a touch and to excessively slight continued pressure, even from a loop of soft thread weighing only the one-sixteenth of a grain (4.05 mg.); and there is reason to believe that the rather thick and stiff petioles of *Clematis flammula* are sensitive to even much less weight if spread over a wide surface. The petioles always bend towards the side which is pressed or touched, at different rates in different species, sometimes within a few minutes, but generally after a much longer period. After temporary contact with any object, the petiole continues to bend for a considerable time; afterwards it slowly becomes straight again and can then re-act. A petiole excited by an extremely slight weight sometimes bends a little and then becomes accustomed to the stimulus and either bends no more or becomes straight again, the weight still remaining suspended. Petioles which have clasped an object for some little time cannot recover their original position. After remaining clasped for two or three days, they generally increase much in thickness either throughout their whole diameter or on one side alone; they subsequently become stronger and more woody, sometimes to a wonderful degree; and in some cases they acquire an internal structure like that of the stem or axis.

C. glandulosa.—The thin upper internodes revolve, moving against the course of the sun, precisely like those of a true twiner, at an average rate, judging from three revolutions, of 3 hrs. 48 m. The leading shoot immediately twined round a stick placed near it; but, after making an open spire of only one turn and a half, it ascended for a short space straight, and then reversed its course and wound two turns in an opposite direction. This was rendered possible by the straight piece between the opposed spires having become rigid. The simple, broad, ovate leaves of this tropical species, with their short thick petioles, seem but ill-fitted for any movement; and whilst twining up a vertical stick, no use is made of them. Nevertheless, if the footstalk of a young leaf be rubbed with a thin twig a few times on any side, it will in the course of a few hours bend to that side; afterwards becoming straight again. The under side seemed to be the most sensitive; but the sensitiveness or irritability is slight compared to that which we shall meet with in some of the following species; thus, a loop of string, weighing 1.64 grain (106.2 mg.) and hanging for some days on a young footstalk, produced a scarcely perceptible effect. A sketch is here given of two young leaves which had naturally caught hold of two thin branches. A forked twig placed so as to press lightly on the under side of a young footstalk caused it, in 12 hrs., to bend greatly, and ultimately to such an extent that the leaf passed to the opposite side of the stem; the forked stick having been removed, the leaf slowly recovered its former position.

The young leaves spontaneously and gradually change their position: when first developed the petioles are upturned and parallel to the stem; they then slowly bend downwards, remaining for a short time at right angles to the stem, and then become so much arched downwards that the blade of the leaf points to the ground with its tip curled inwards, so that the whole petiole and lead together form a hook. They are thus enabled to catch hold of any twig with which they may be brought into contact by the revolving movement of the internodes. If this does not happen, they retain their hooked shape for a considerable time, and then bending upwards reassume their original upturned position, which is preserved ever afterwards. The petioles which have clasped any object soon become much thickened and strengthened, as may be seen in the drawing.

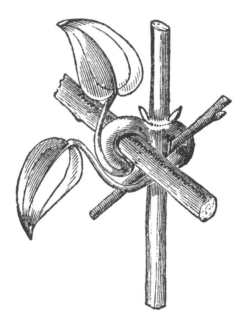

Clematis glandulosa. With two young leaves clasping two twigs, with the clasping portions thickened.

Clematis montana.—The long, thin petioles of the leaves, whilst young, are sensitive, and when lightly rubbed bend to the rubbed side, subsequently becoming straight. They are far more sensitive than the petioles of *C. glandulosa*; for a loop of thread weighing a quarter of a grain (16.2 mg.) caused them to bend; a loop weighing only one-eighth of a grain (8.1 mg.) sometimes acted and sometimes did not act. The sensitiveness extends from the blade of the leaf to the stem. I may here state that I ascertained in all cases the weights of the string and thread used by carefully weighing 50 inches in a chemical balance, and then cutting off measured lengths. The main petiole carries three leaflets; but their short, sub-petioles are not sensitive. A young, inclined shoot (the plant being in the greenhouse) made a large circle opposed to the course of the sun in 4 hrs. 20 m., but the next day, being very cold, the time was 5 hrs. 10m. A stick placed near a revolving stem was soon struck by the petioles which stand out at right angles, and the revolving movement was thus arrested. The petioles then began, being excited by the contact, to slowly wind round the stick. When the stick was thin, a petiole sometimes wound twice round it. The opposite leaf was in no way affected. The attitude assumed by the stem after the petiole had clasped the stick, was that of a man standing by a column, who throws his arm horizontally round it.

Sɥᵈ Edwards del.   Pub by T Curtis, St Geo: Crescent July 1.1805.   F. Sansom sculp

Cobaea scandens. Hand-colored engraving drawn by Sydenham Edwards,
from *The Botanical Magazine* 22: 851.

Cobaea scandens
CUP AND SAUCER VINE

——— ...... ———

POLEMONIACEAE—PHLOX FAMILY

CLIMBING PLANTS

Cobaea are vines and lianas of the Phlox family, with about twenty species found from tropical Mexico to northern South America. Several are prized in the horticultural trade, but perhaps none more than cup and saucer vine, *C. scandens*, an elegant and vigorous climber of southern Mexico. The common name is a nod to its fragrant espresso cup–sized flower with its green calyx "saucer."

In July 1863, as Darwin's fascination with climbing plants expanded, he asked Joseph Hooker at Kew for seeds of tendril bearing plants, including *Cobaea*. Hooker wrote him, encouragingly, "Your observations on Tendrils &c are most curious and novel, and I am delighted that you are going on with them—you are 'facile princeps' [clearly the leader] of observers."[38] Hooker told his friend he was looking out for climbers to send him, and Darwin received a consignment of *Cobaea* and others soon after.

Darwin declared this species an "excellently constructed climber," describing its long, slender tendrils with branches that terminate in tiny double grappling hooks. But he was most impressed with the (relative) rapidity of the tendril's circumnutation, revolving in a circular or elliptical motion. He clocked one complete rotation of the tendrils at 1 hour 15 minutes, garnering a double exclamation in his experimental notebook.[39] He later reported this feat more matter-of-factly in *Climbing Plants*, remarking on the ability of the tendrils to catch hold of branches for support.

From: *The Movements and Habits of Climbing Plants*
(2nd ed., 1875)

Cobaea scandens—This is an excellently constructed climber. The tendrils on a fine plant were eleven inches long, with the petiole bearing two pairs of leaflets, only two and a half inches in length. They revolve more rapidly and vigorously than those of any other tendril-bearer observed by me, with the exception of one kind of *Passiflora*. Three large, nearly circular sweeps, directed against the sun, were completed, each in 1 hr. 15 m.; and two other circles in 1 hr. 20 m. and 1 hr. 23 m. Sometimes a tendril travels in a much inclined position, and sometimes nearly upright. ... The long, straight, tapering main stem of the tendril of the *Cobaea* bears alternate branches; and each branch is several times divided, with the finer branches as thin as very thin bristles and extremely flexible, so that they are blown about by a breath of air; yet they are strong and highly elastic. The extremity of each branch is a little flattened and terminates in a minute double (though sometimes single) hook, formed of a hard, translucent, woody substance, and as sharp as the finest needle. On a tendril which was eleven inches long, I counted ninety-four of these beautifully constructed little hooks. They readily catch soft wood, or gloves, or the skin of the naked hand. With the exception of these hardened hooks, and of the basal part of the central stem, every part of every branchlet is highly sensitive on all sides to a slight touch and bends in a few minutes towards the touched side. By lightly rubbing several sub-branches on opposite sides, the whole tendril rapidly assumed an extraordinarily crooked shape. These movements from contact do not interfere with the ordinary revolving movement. The branches, after becoming greatly curved from being touched, straighten themselves at a quicker rate than in almost any other tendril seen by me, namely, in between half an hour and an hour. After the tendril has caught any object, spiral contraction likewise begins after an unusually short interval of time, namely, in about twelve hours.

Before the tendril is mature, the terminal branchlets cohere, and the hooks are curled closely inwards. At this period no part is sensitive to a touch; but as soon as the branches diverge and the hooks stand out, full sensitiveness is acquired. It is a singular circumstance that immature tendrils revolve at their full velocity before they become sensitive, but in a useless manner, as in this state they can catch nothing. This want of perfect co-adaptation, though only for a short time, between the structure and the functions of a climbing-plant is a rare event. A tendril, as soon as it is ready to act, stands, together with the supporting petiole, vertically upwards. The leaflets borne by the petiole are at this time quite small, and the extremity of the growing stem is bent to one side so as to be out of the way of the revolving tendril, which sweeps large circles directly over head. The tendrils thus revolve in a position well adapted for catching objects standing above; and by this means, the ascent of the plant is favoured. If no object is caught, the leaf with its tendril bends downwards and ultimately assumes a horizontal position. An open space is thus left for the next succeeding and younger tendril to stand vertically upwards and to revolve freely. As soon as an old tendril bends downwards, it loses all power of movement, and contracts spirally into an entangled mass. Although the tendrils revolve with unusual rapidity, the movement lasts for only a short time. In a plant placed in the hot-house and growing vigorously, a tendril revolved for not longer than 36 hours, counting from the period when it first became sensitive; but during this period it probably made at least 27 revolutions.

When a revolving tendril strikes against a stick, the branches quickly bend round and clasp it. The little hooks here play an important part, as they prevent the branches from being dragged away by the rapid revolving movement, before they have had time to clasp the stick securely.

The perfect manner in which the branches arranged themselves, creeping like rootlets over every inequality of the surface and into any deep crevice, is a pretty sight; for it is perhaps more effectually performed by this than by any other species. The action is certainly more conspicuous, as the upper surfaces of the main stem, as well as of every branch to the extreme hooks, are angular and green, whilst the lower surfaces are rounded and purple. I was led to infer, as in former cases, that a less amount of light guided these movements of the branches of the tendrils. I made many trials with black and white cards and glass tubes to prove it but failed from various causes; yet these trials countenanced the belief. As a tendril consists of a leaf split into numerous segments, there is nothing surprising in all the segments turning their upper surfaces towards the light as soon as the tendril is caught and the revolving movement is arrested. But this will not account for the whole movement, for the segments actually bend or curve to the dark side besides turning round on their axes so that their upper surfaces may face the light.

When the *Cobaea* grows in the open air, the wind must aid the extremely flexible tendrils in seizing a support, for I found that a mere breath sufficed to cause the extreme branches to catch hold by their hooks of twigs, which they could not have reached by the revolving movement. It might have been thought that a tendril, thus hooked by the extremity of a single branch, could not have fairly grasped its support.

But several times I watched cases like the following: a tendril caught a thin stick by the hooks of one of its two extreme branches; though thus held by the tip, it still tried to revolve, bowing itself to all sides, and by this movement, the other extreme branch soon caught the stick. The first branch then loosed itself, and, arranging its hooks, again caught hold. After a time, from the continued movement of the tendril, the hooks of a third branch caught hold. No other branches, as the tendril then stood, could possibly have touched the stick. But before long the upper part of the main stem began to contract into an open spire. It thus dragged the shoot which bore the tendril towards the stick; and as the tendril continually tried to revolve, a fourth branch was brought into contact. And lastly, from the spiral contraction travelling down both the main stem and the branches, all of them, one after another, were ultimately brought into contact with the stick. They then wound themselves round it and round one another, until the whole tendril was tied together in an inextricable knot. The tendrils, though at first quite flexible, after having clasped a support for a time, become more rigid and stronger than they were at first. Thus the plant is secured to its support in a perfect manner.

Coryanthes maculata var. *punctata*. Lithograph by John Nugent Fitch,
from Robert Warner, *The Orchid Album*.

```
             Coryanthes
        BUCKET ORCHID
            ——— ······ ———
    ORCHIDACEAE—ORCHID FAMILY
```

ORCHIDS, FORMS OF FLOWERS, POLLINATION

Coryanthes is a genus of nearly fifty Neotropical species, pollinated by jewel-like orchid bees that have coevolved with them. Orchid bees are in the subfamily Euglossini, closely related to Bombini, the bumblebees or, in Darwin's day, humble-bees. They are spectacular metallic-iridescent insects, and the males have an enlarged, hollow tibia on each hind leg. These serve as storage vessels for aromatic oils produced by the plants, which the males use to court female bees.[40] The inverted bucket-shaped lip of the flower, the labellum, gives the group its common name and stores the liquid which contains the oils. When a bee falls into the bucket, the liquid (mostly water) wets its wings, preventing it from flying away and compelling it to climb out of the bucket through a narrow spout. As it squeezes through, the orchid's pollinia become attached to its thorax, ready to be transferred to another flower.

Darwin's research on *Coryanthes* was facilitated by his correspondent in Trinidad, Hermann Crüger, then director of the Royal Botanic Gardens in Port-of-Spain. Excited as always by the intricate structure and function of orchid flower parts, he enthused to Asa Gray, "Cruger's account of *Coryanthes* and the use of the bucket-like labellum full of water beats everything. I suspect the bees being well wetted flattens hairs and allows viscid disc to adhere."[41] Aware of the "inexhaustible number of contrivances" of orchids, he considered *Coryanthes* a case of extraordinary adaption. Crüger sent him bees and flowers preserved in alcohol in 1864,[42] and wrote his own treatise on the genus in 1864. Darwin first discussed *Coryanthes* in the fourth edition of *Origin*, where he marveled at this "acme of perfect adaptation,"[43] then went into more detail in his orchid book.

[From: *On the Origin of Species* (4th ed., 1866)]

[*Coryanthes*] has its labellum or lower lip hollowed out into a great bucket, into which drops of almost pure water, not nectar, continually fall from two secreting horns which stand above it; and when the bucket is half full, the water overflows by a spout on one side. The basal part of the labellum curves over the bucket and is itself hollowed out into a sort of chamber with two lateral entrances, within which and outside there are some curious fleshy ridges.

The most ingenious man, if he had not witnessed what takes place, could never have imagined what purpose all these parts served. But Dr. Crüger saw crowds of large humble-bees visiting the gigantic flowers of this orchid in the early morning, and they came, not to suck nectar, but to gnaw off the ridges above the bucket; in doing this they frequently pushed each other into the bucket, and thus their wings were wetted, so that they could not fly out, but had to crawl out through the passage formed by the spout or overflow. Dr. Crüger has seen a "continual procession" of bees thus crawling out of their involuntary bath. The passage is narrow, and is roofed over by the column, so that a bee, in forcing its way out, first rubs its back against the viscid stigma and then against the viscid glands of the pollen-masses. The pollen-masses are thus glued to the back of the bee which first happens to crawl through the passage of a lately expanded flower and are thus carried away. Dr. Crüger sent me a flower in spirits of wine, with a bee which he had killed before it had quite crawled out of the passage with a pollen-mass fastened to its back. When the bee, thus provided, flies to another flower, or to the same flower a second time, and is pushed by its comrades into the bucket and then crawls out by the passage, the pollen-mass necessarily comes first into contact with the viscid stigma and adheres to it, and the flower is fertilised. Now at last we see the full use of the water-secreting horns, of the bucket with its spout, and of the shape of every part of the flower!

[
From: *The Various Contrivances by Which Orchids are Fertilised by Insects* (2nd ed., 1877)
]

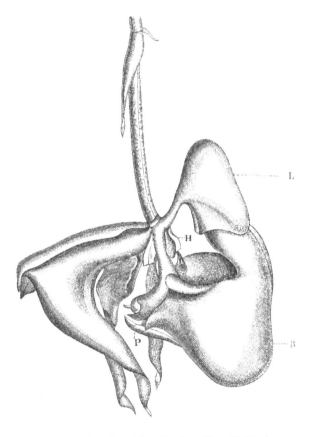

Coryanthes speciosa. Copied from Lindley's *Vegetable Kingdom* (1846–1853) L. labellum. B. bucket of the labellum. H. fluid-secreting appendages. P. spout of bucket, over-arched by the end of the column, bearing the anther and stigma.

Coryanthes.— The flowers are very large and hang downwards. The distal portion of the labellum (L) in the woodcut is converted into a large bucket (B). Two appendages (H), arising from the narrowed base of the labellum, stand directly over the bucket and secrete so much fluid that drops may be seen falling into it. This fluid is limpid and so slightly sweet that it does not deserve to be called nectar, though evidently of the same nature; nor does it serve to attract insects. When the bucket is full the fluid overflows by the spout. This spout is closely over-arched by the end of the column, which bears the stigma and pollen-masses in such a position that an insect forcing its way out of the bucket through this passage would first brush with its back against the stigma and afterwards against the viscid discs of the pollinia, and thus remove them. We are now prepared to hear what Dr. Crüger says about the fertilisation of an allied species, the *C. macrantha*, the labellum of which is provided with crests. I may premise that he sent me specimens of the bees which he saw gnawing these crests, and they belong to the genus *Euglossa*. Dr. Crüger states that these bees may be "seen in great numbers disputing with each other for a place on the edge of the hypochil (i.e. the basal part of the labellum). Partly by this contest, partly perhaps intoxicated by the matter they are indulging in, they tumble down into the 'bucket,' half-full of a fluid secreted by organs situated at the base of the column. They then crawl along in the water towards the anterior side of the bucket, where there is a passage for them between the opening of this and the column. If one is early on the look-out, as these Hymenopterae are early risers, one can see in every flower how fecundation is performed. The humble-bee, in forcing its way out of its involuntary bath, has to exert itself considerably, as the mouth of the epichil (i.e. the distal part of the labellum) and the face of the column fit together exactly, and are very stiff and elastic. The first bee, then, which is immersed will have the gland of the pollen-mass glued to its back. The insect then generally gets through the passage, and comes out with this peculiar appendage, to return nearly immediately to its feast, when it is generally precipitated a second time into the bucket, passing out through the same opening, and so inserting the pollen-masses into the stigma while it forces its way out and thereby impregnating either the same or some other flower. I have often seen this; and sometimes there are so many of these humble-bees assembled that there is a continual procession of them through the passage specified."

There cannot be the least doubt that the fertilisation of the flower absolutely depends on insects crawling out through the passage formed by the extremity of the labellum and the over-arching column. If the large distal portion of the labellum or bucket had been dry, the bees could easily have escaped by flying away. Therefore we must believe that the fluid is secreted by the appendages in such extraordinary quantity and is collected in the bucket, not as a palatable attraction for the bees, as these are known to gnaw the labellum, but for the sake of wetting their wings and thus compelling them to crawl out through the passage.

Cyclamen europaeum (= *purpurascens*). Watercolor by Elizabeth Pieth Schmitz, *Botanical Manuscript*.

Cyclamen

CYCLAMEN

———— ...... ————

PRIMULACEAE—PRIMROSE FAMILY

CROSS AND SELF-FERTILIZATION, PLANT MOVEMENT

Cyclamen numbers about twenty-three species ranging from Europe to the Middle East and north Africa. The name is derived from the Greek *kyklos*, circular, referring to the disk-like shape of the tuber. The plants are prized for their boldly patterned leaves and dramatic flowers, nodding with backswept petals. After pollination, as the fruit develops, nearly all *Cyclamen* species exhibit a curious change: the upright flower stem (peduncle) lengthens as it goes into a slow-motion, spiraling nose-dive, carrying the seed capsule down under the leaf litter and, in some cases, right into the ground. This might seem like a kind of "self-sowing," but it is thought to function more as a way to provide ants ready access to the ripening capsule—*Cyclamen* seeds are myrmecochorous, meaning they are dispersed by ground-dwelling ants, which are rewarded for their efforts with a nutritious morsel called an elaiosome that's attached to each seed. Darwin was aware of the curious descent of *Cyclamen* seed capsules—his grandfather Erasmus Darwin commented upon it in his epic poem *The Loves of the Plants*.*

---

* In *The Loves of the Plants*, Part II of the 1791 poem *The Botanic Garden*, the polymathic physician, inventor, and poet Erasmus Darwin fancifully interpreted and anthropomorphized the botany of his day in heroic couplets that drew heavily upon classical mythological allegory. In "Canto III," the burying of seed capsules in cyclamen is imagined as the burial of a deceased infant, to be born again: (cont.)

Charles mainly studied Persian cyclamen, *Cyclamen persicum*, the one species that does not coil its peduncles, though they do curve as the seed capsules are lowered to the ground. He grew this species in his greenhouse where he was able to observe and experiment on several levels. He crossed and self-pollinated flowers and then raised offspring from the resulting seeds over several years, noting that cross fertilization yielded robust plants, whereas self-fertilized plants produced "miserable specimens."

$$\Big[\ \textbf{From: } \textit{The Effects of Cross and Self Fertilisation in} \\ \textit{the Vegetable Kingdom } \textbf{(2nd ed., 1878)}\ \Big]$$

Ten flowers crossed with pollen from plants known to be distinct seed-lings, yielded nine capsules, containing on an average 34.2 seeds, with a maximum of seventy-seven in one. Ten flowers self-fertilised yielded eight capsules, containing on an average only 13.1 seeds, with a maximum of twenty-five in one. This gives a ratio of 100 to 38 for the average number of seeds per capsule for the crossed and self-fertilised flowers. The flowers hang downwards, and as the stigmas stand close beneath the anthers, it might have been expected that pollen would have fallen on them and that they would have been spontaneously self-fertilised; but these covered-up plants did not produce a single capsule. On some other occasions, uncovered plants in the same greenhouse produced plenty of capsules,

---

(cont.)
The gentle CYCLAMEN with dewy eye
Breathes o'er her lifeless babe the parting sigh;
And, bending low to earth, with pious hands
Inhumes her dear Departed in the sands.
"Sweet Nursling! withering in thy tender hour,
"Oh sleep," she cries, "and rise a fairer flower!

and I suppose that the flowers had been visited by bees, which could hardly fail to carry pollen from plant to plant.

The seeds obtained in the manner just described were placed on sand, and after germinating were planted in pairs—three crossed and three self-fertilised plants on the opposite sides of four pots. When the leaves were 2 or 3 inches in length, including the foot-stalks, the seedlings on both sides were equal. In the course of a month or two, the crossed plants began to show a slight superiority over the self-fertilised, which steadily increased; and the crossed flowered in all four pots some weeks before, and much more profusely than the self-fertilised. The two tallest flower-stems on the crossed plants in each pot were now measured, and the average height of the eight stems was 9.49 inches. After a considerable interval of time, the self-fertilised plants flowered, and several of their flower-stems were roughly measured, and their average height was a little under 7.5 inches; so that the flower-stems on the crossed plants to those on the self-fertilised were at least as 100 to 79. ...

These plants were left uncovered in the greenhouse; and the twelve crossed plants produced forty capsules, whilst the twelve self-fertilised plants produced only five; or as 100 to 12. But this difference does not give a just idea of the relative fertility of the two lots. I counted the seeds in one of the finest capsules on the crossed plants, and it contained seventy-three; whilst the finest of the five capsules produced by the self-fertilised plants contained only thirty-five good seeds. In the other four capsules most of the seeds were barely half as large as those in the crossed capsules. In the following year, the crossed plants again bore many flowers before the self-fertilised bore a single one. The self-fertilised plants were miserable specimens, whilst the crossed ones looked very vigorous.

Darwin's observations on this cyclamen's leaf movement revealed a zigzag pattern, rising in the evening and falling in the morning. He watched, fascinated, as the peduncles of the fertilized flowers slowly lengthened and bowed their developing seed capsule to the soil in a graceful curve. His experiments with potted plants, brought in and out of a dark cupboard, confirmed that it was not gravity that pulled the capsules down, but a distinct form of plant movement, a modified form of the common circular motion (circumnutation) he dubbed "apheliotropism"—movement away from the sun.

[From: *The Power of Movement in Plants* (1880)]

Whilst this plant is in flower, the peduncles stand upright, but their upper-most part is hooked so that the flower itself hangs downwards. As soon as the pods begin to swell, the peduncles increase much in length and slowly curve downwards, but the short, upper, hooked part straightens itself. Ulti-mately the pods reach the ground, and if this is covered with moss or dead leaves, they bury themselves. We have often seen saucer-like depressions formed by the pods in damp sand or sawdust; and one pod (.3 of inch in diameter) buried itself in sawdust for three-quarters of its length. We shall have occasion hereafter to consider the object gained by this burying process. The peduncles can change the direction of their curvature, for if a pot, with plants having their peduncles already bowed downwards, be placed horizontally, they slowly bend at right angles to their former direc-tion towards the centre of the earth. We therefore at first attributed the movement to geotropism; but a pot which had lain horizontally with the pods all pointing to the ground, was reversed, being still kept horizontal, so that the pods now pointed directly upwards; it was then placed in a dark cupboard, but the pods still pointed upwards after four days and nights. The pot, in the same position, was next brought back into the light, and after two days there was some bending downwards of the peduncles, and on the fourth day two of them pointed to the centre of the earth, as did the others after an additional day or two. Another plant, in a pot which had always stood upright, was left in the dark cupboard for six days; it bore 3 peduncles, and only one became within this time at all bowed downwards, and that doubtfully. The weight, therefore, of the pods is not the cause of the bending down. This pot was then brought back into the light, and after three days, the peduncles were considerably bowed downwards. We are thus led to infer that the downward curvature is due to apheliotropism; though more trials ought to have been made.

In order to observe the nature of this movement, a peduncle bearing a large pod which had reached and rested on the ground, was lifted a little up and secured to a stick. A filament was fixed across the pod with a mark beneath, and its movement, greatly magnified, was traced on a horizontal glass during 67 h. The plant was illuminated during the day from above. A copy of the tracing is given [here]; and there can be no doubt that the descending movement is one of modified circumnutation, but on an extremely small scale. ... Considering the great length and thinness of the peduncles and the lightness of the pods, we may conclude that they would not be able to excavate saucer-like depressions in sand or sawdust, or bury themselves in moss, etc., unless they were aided by their continued rocking or circumnutating movement.

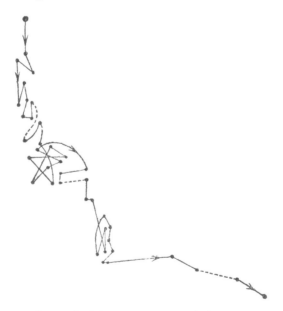

Cyclamen persicum: downward apheliotropic movement of a flower-peduncle, traced on a horizontal glass from 1 P.M. Feb. 18th to 8 A.M. 21st.

Cypripedium calceolus   *Ladies Slipper*

Cypripedium calceolus. Watercolor by Elizabeth Wharton, *British Flowers.*

ORCHIDS, FORMS OF FLOWERS, POLLINATION

The slipper orchids, *Cypripedium*, belong to a group that includes the genera *Paphiopedilum*, *Phragmipedium*, and *Selenipedium*, with structures distinctly different from other orchids. In Darwin's time, they were all placed in the genus *Cypripedium* by botanist John Lindley. Darwin wrote about their unusual flower structure in the first edition of *Orchids* and then pursued further research and corrected many of his initial ideas in the second edition of 1877. His initial research involved the tropical species now known as *Paphiopedilum insigne*, one he could grow in his greenhouse. It all began in 1861, when he requested Daniel Oliver, horticulturist at the Royal Botanic Gardens, Kew, to obtain plants for him, and further research developed when he wrote to Asa Gray at Harvard to observe related North American species and look for insect pollinators. "Could you cover up a plant with net and leave one uncovered," he asked Gray, "*if it be one which sets seeds*, and see whether the *protected* one sets seeds, and whether the pollen of the two after interval of time are in the same state."[44]

Gray examined the pollination mechanisms and concluded that, for the North American species, the account given by Darwin was incorrect. Darwin had suggested that *Cypripedium* must be pollinated by an insect inserting its proboscis into one of the two lateral entrances at the base of the labellum, directly over one of the two lateral anthers, and thus either placing the pollen onto the flower's own stigma or carrying it away to another flower. Gray, however, suggested that the flowers were pollinated by insects entering the labellum through the large opening on the upper surface, and crawling out through one of the two smaller lateral openings.

Darwin did not receive Gray's observations in time to be included in the first edition of his orchid book, but in the second edition, he cited Gray's descriptions of North American species and quoted his observation that these orchids are "excellently adapted to brush off the pollen from an insect's head or back." Gray sent Darwin three slipper orchid species (*C. acaule*, *spectabile*, and *arietinum*) and instructed him to "pot and keep them cold till spring, and then let grow; they will surely flower."[45] A year later, Darwin confirmed that insects visit the flowers as Gray suggested, and he noted that his colleagues Herman Müller and Federico Delpino observed pollinators in another species, *Cypripedium calceolus*, doing much the same thing. He acknowledged his error to entomologist Roland Trimen, in South Africa, writing, "By the bye, I believe I have blundered on *Cypripedium*: Asa Gray suggested that small insects enter by the toe and crawl out by the lateral windows.— I put in a small bee and it did so and came out with its back smeared with pollen; I caught him and put him in again, and again he crawled out by the window: I cut open the flower and found the stigma smeared with pollen!"[46]

Everyone makes mistakes, and Darwin was dogged in his efforts to correct his. He analyzed *Cypripedium* inside and out, as indicated in letters to clergyman and orchid enthusiast Arthur Rawson, who offered to "lend" him some Yellow Lady's Slippers (*C. pubescens*). Darwin replied, "I certainly should be very glad of the loan of the *Cypripedium*. But do you understand that a 'loan' means that I should probably cut up and mutilate all or nearly all the flowers; without doing this the flowers would be of no use to me." He continued, jokingly: "Are you prepared to be so generous a martyr-florist? If so, I will gratefully send for plant, whenever I hear that it is ready."[47]

Rawson was happy to sacrifice his orchid flowers for a good cause. Two months later, Darwin sent him a note of thanks and reported the results. "The contrivance by which insects, after they have entered the labellum by the mouth, are forced to crawl out by one of the small lateral passages and thus get smeared with the viscid pollen, is very curious; it is exactly the same principle on which traps are made to catch insects in kitchen, namely the edge of the mouth, or large opening into the labellum being bent inwards so that the insect instead of getting out falls back."[48]

[**From: *The Various Contrivances by Which Orchids are Fertilised by Insects* (2nd ed., 1877)**]

We have now arrived at Lindley's last and seventh tribe, including, according to most botanists, only a single genus, *Cypripedium*, which differs from all other Orchids far more than any other two of these do from one another. An enormous amount of extinction must have swept away a multitude of intermediate forms and has left this single genus, now widely distributed, as a record of a former and more simple state of the great Orchidean Order. *Cypripedium* possesses no rostellum; for all three stigmas are fully developed, though confluent. The single anther, which is present in all other Orchids, is here rudimentary and is represented by a singular shield-like projecting body, deeply notched or hollowed out on its lower margin. There are two fertile anthers which belong to an inner whorl, represented in ordinary Orchids by various rudiments. The grains of pollen are not united together by threes or fours, as in so many other genera, nor are they tied together by elastic threads, nor furnished with a caudicle, nor cemented into waxy masses. The labellum is of large size and is a compound organ as in all other Orchids.

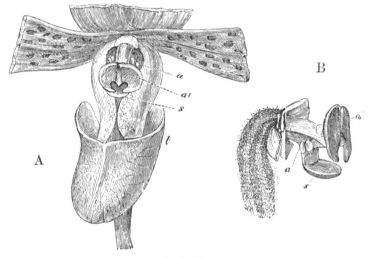

Cypripedium.
a. anther. *a'.* rudimentary, shield-like anther. *s.* stigma. *l.* labellum.

A. Flower viewed from above, with the sepals and petals, excepting the labellum, partly cut off. The labellum has been slightly depressed, so that the dorsal surface of the stigma is exposed; the edges of the labellum have thus become a little separated and the toe or extremity stands lower than is natural.

B. Side view of column, with all the sepals and petals removed.

The basal part of the labellum is folded round the short column, so that its edges nearly meet along the dorsal surface; and the broad extremity is folded over in a peculiar manner, forming a sort of shoe, which closes up the end of the flower. Hence arises the English name of Ladies' slipper. The overarching edges of the labellum are inflected or sometimes only smooth and polished internally; and this is of much importance, as it prevents insects which have once entered the labellum from escaping through the great opening in the upper surface. In the position in which the flower grows, as here represented, the dorsal surface of the column is uppermost. The stigmatic surface is slightly protuberant and is not adhesive; it stands nearly parallel to the lower surface of the labellum. With a flower in its natural state, the margin of the dorsal surface of the stigma can be barely distinguished between the edges of the labellum and through the notch in the rudimentary, shield-like anther (a'); but in the drawing (s, fig. A) the margin of the stigma has been brought outside the edges of the depressed labellum, and the toe is a little bent downwards, so that the flower is represented as rather more open than it really is. The edges of the pollen-masses of the two lateral anthers (a) can be seen through the two small orifices or open spaces in the labellum (fig. A) on each side, close to the column. These two orifices are essential for the fertilisation of the flower. ...

I have never been able to detect nectar within the labellum ... The inner surface of the labellum, however, in those species which I examined, is clothed with hairs, the tips of which secrete little drops of slightly viscid fluid. And these, if sweet or nutritious, would suffice to attract insects. The fluid when dried forms a brittle crust on the summits of the hairs. Whatever the attraction may be, it is certain that small bees frequently enter the labellum.

Formerly I supposed that insects alighted on the labellum and inserted their proboscides through either of the orifices close to the anthers; for I found that when a bristle was thus inserted the glutinous pollen adhered to it and could afterwards be left on the stigma; but this latter part of the operation was not well effected. After the publication of my book, Professor Asa Gray wrote to me that he was convinced from an examination of several American species that the flowers were fertilised by small insects entering the labellum through the large opening on the upper surface and crawling out by one of the two small orifices close to the anthers and stigma. Accordingly, I first introduced some flies into the labellum of C. pubescens, through the large upper opening, but they were either too large or too stupid, and did not crawl out properly. I then caught and placed within the labellum a very small bee which seemed of about the right size, namely, Andrena parvula, and this by a strange chance proved, as we shall presently see, to belong to the genus on which in a state of nature the fertilisation of C. calceolus depends. The bee vainly endeavoured to crawl out again the same way by which it had entered, but always fell backwards, owing to the margins being inflected. The labellum thus acts like one of those conical traps with the edges turned inwards, which are sold to catch beetles and cockroaches in the London kitchens. It could not creep out through the slit between the folded edges of the basal part of the labellum, as the elongated, triangular, rudimentary stamen here closes the passage. Ultimately it forced its way out through one of the small orifices close to one of the anthers and was found when caught to be smeared with the glutinous pollen. I then put the same bee back into the labellum; and again it crawled out through one of the small orifices, always covered with pollen. I repeated the operation five times, always with the same result. I afterwards cut away the labellum, so as to examine the stigma, and found its whole surface covered with pollen. It should be noticed that an insect in making its escape must first brush past the stigma and afterwards one of the anthers, so that it cannot leave pollen on the stigma until, being already smeared with pollen from one flower, it enters another; and thus there will be a good chance of cross-fertilisation between two distinct plants. Delpino with much sagacity foresaw that some insect would be discovered to act in this manner; for he argued that if an insect were to insert its proboscis, as I had supposed, from the outside through one of the small orifices close to one of the anthers, the stigma would be liable to be fertilised by the plant's own pollen: and in this he did not believe, from having confidence in what I have often insisted on—namely, that all the contrivances for fertilisation are arranged so that the stigma shall receive pollen from a distinct flower or plant. But these speculations are now all superfluous; for, owing to the admirable observations of Dr. H. Müller, we know that *Cypripedium calceolus* in a state of nature is fertilised in the manner just described, by bees belonging to five species of *Andrena*.

Thus, the use of all the parts of the flower, namely, the inflected edges, or the polished inner sides of the labellum,—the two orifices and their position close to the anthers and stigma, the large size of the medial rudimentary stamen,—are rendered intelligible. An insect which enters the labellum is thus compelled to crawl out by one of the two narrow passages, on the sides of which the pollen-masses and stigma are placed.

Admiral Vernon.

G. D. Ehret. pinxit.
1756.

Dianthus caryophyllus. Water and bodycolor on vellum by Georg Dionysius Ehret,
in *Flowers, Moths, Butterflies and Shells*.

Dianthus

CARNATION

——— ...... ———

CARYOPHYLLACEAE—
CARNATION FAMILY

CROSS AND SELF-FERTILIZATION

The genus *Dianthus*, derived from Greek terms meaning "divine flower" for its beauty and fragrance, includes around 300 species native mainly to Europe and Asia. Several *Dianthus* species were prized by the ancients, and two of them played a starring role in horticultural history when, in 1717, the renowned English horticulturist Thomas Fairchild produced what may be the first experimental hybrid plant by crossing *D. caryophyllus* (Carnation) with *D. bartatus* (Sweet William).[49] "Fairchild's Mule," as the hybrid was called, was widely celebrated at the time—Erasmus Darwin, Charles's grandfather, waxed poetic over the feat in *The Loves of the Plants*. At a time when the nature of reproduction in plants and Carl Linnaeus's sex-based botanical classification system were hotly debated, Erasmus and others cited Fairchild's Mule as evidence in support of the Swedish savant. The sterility of "vegetable mules," Erasmus noted, "supply an irrefragable argument in favour of the sexual system of botany."[50]

Carnations were especially popular in Darwin's era, and their popularity continues—today the International Dianthus Register of the Royal Horticultural Society lists over 30,000 cultivars. Darwin began his interest in crossbreeding of carnation flowers in 1855, requesting seeds from his college mentor John Stevens Henslow to try experiments on hybridization. He asked for seeds of "wild" *Dianthus caryophyllus* but it was hardly a wild plant, with its exact range in nature unknown since it had been cultivated for more than 2000 years.

Soon after he studied pollination in orchid flowers, Darwin began systematic self- and cross-fertilization experiments on many genera of plants. It all started with an observation he first made with *Linaria* (see p. 185) when he planted, side-by-side, beds of seedlings produced by selfing versus crossing. "To my surprise, the crossed plants when fully grown were plainly taller and more vigorous than the self-fertilised ones," he marveled. "During the next year, I raised for the same purpose as before two large beds close together of self-fertilised and crossed seedlings from the carnation, *Dianthus caryophyllus*." The result was the same. His attention "now thoroughly aroused," he set about making thousands of crosses with dozens of plant species, resulting in the publication of *The Effects of Cross and Self Fertilisation in the Vegetable Kingdom* in 1876.[51]

Dianthus flowers are termed "protandrous" (Darwin's "proterandrous"), with the anthers maturing as much as a week before the stigmas become receptive, and their fragrance is not only appealing to people but certainly to the many insects, including humble-bees (a.k.a., bumblebees) that pollinate them with abandon. Darwin's experiments confirmed the advantage of cross-pollination between *Dianthus* flowers of different plants rather than on the same plant. He crossed and selfed plants and collected seeds over four generations, measuring the height and weight and counting the number of seeds produced by each plant in each successive generation and observing the flower color and patterns yielded by the crosses. His results confirmed that horticulturalists should rely on cross-pollination—outcrossing—to obtain the most robust and fertile plants.

From: *The Effects of Cross and Self Fertilisation in the Vegetable Kingdom* (2nd ed., 1878)

Dianthus caryophyllus. The common carnation is strongly proterandrous, and therefore depends to a large extent upon insects for fertilisation. I have seen only humble-bees visiting the flowers, but I dare say other insects likewise do so. It is notorious that if pure seed is desired, the greatest care is necessary to prevent the varieties which grow in the same garden from intercrossing. The pollen is generally shed and lost before the two stigmas in the same flower diverge and are ready to be fertilised. I was therefore often forced to use for self-fertilisation pollen from the same plant instead of from the same flower. ...

Several single-flowered carnations were planted in good soil and were all covered with a net. Eight flowers were crossed with pollen from a distinct plant and yielded six capsules, containing on an average 88.6 seeds, with a maximum in one of 112 seeds. Eight other flowers were self-fertilised in the manner above described, and yielded seven capsules containing on an average 82 seeds, with a maximum in one of 112 seeds. So that there was very little difference in the number of seeds produced by cross-fertilisation and self-fertilisation, viz., as 100 to 92. As these plants were covered by a net, they produced spontaneously only a few capsules containing any seeds, and these few may perhaps be attributed to the action of Thrips and other minute insects which haunt the flowers. A large majority of the spontaneously self-fertilised capsules produced by several plants contained no seeds, or only a single one. Excluding these latter capsules, I counted the seeds in eighteen of the finest ones, and these contained on an average 18 seeds. ...

Crossed and self-fertilised Plants of the First Generation.—The many seeds obtained from the above crossed and artificially self-fertilised flowers were sown out of doors, and two large beds of seedlings, closely adjoining one another, thus raised. This was the first plant on which I experimented, and I had not then formed any regular scheme of operation. When the two lots were in full flower, I measured roughly a large number of plants, but record only that the crossed were on an average fully 4 inches taller than the self-fertilised. Judging from subsequent measurements, we may assume that the crossed plants were about 28 inches, and the self-fertilised about 24 inches in height; and this will give us a ratio of 100 to 86. Out of a large number of plants, four of the crossed ones flowered before any one of the self-fertilised plants.

Thirty flowers on these crossed plants of the first generation were again crossed with pollen from a distinct plant of the same lot, and yielded twenty-nine capsules, containing on an average 55.62 seeds, with a maximum in one of 110 seeds.

Crossed and self-fertilised Plants of the Second Generation.—The crossed and self-fertilised seeds from the crossed and self-fertilised plants of the last generation were sown on opposite sides of two pots; ... Some flowers on these crossed plants were again crossed with pollen from another plant of the same lot, and some flowers on the self-fertilised plants again self-fertilised; and from the seeds thus obtained the plants of the next generation were raised.

Crossed and self-fertilised Plants of the Third Generation.—The seeds just alluded to were allowed to germinate on bare sand and were planted in pairs on the opposite sides of four pots. When the seedlings were in full flower, the tallest stem on each plant was measured to the base of the calyx. ... In Pot I. the crossed and self-fertilised plants flowered at the same time; but in the other three pots the crossed flowered first. These latter plants also continued flowering much later in the autumn than the self-fertilised.

The average height of the eight crossed plants is here 28.39 inches, and of the eight self-fertilised 28.21; or as 100 to 99. So that there was no difference in height worth speaking of, but in general vigour and luxuriance there was an astonishing difference, as shown by their weights. After the seed-capsules had been gathered, the eight crossed and the eight self-fertilised plants were cut down and weighed; the former weighed 43 ounces, and the latter only 21 ounces; or as 100 to 49. ...

In summary: This plant was experimented on during four generations, in three of which the crossed plants exceeded in height the self-fertilised generally by much more than five per cent.; and we have seen that the offspring from the plants of the third self-fertilised generation crossed by a fresh stock profited in height and fertility to an extraordinary degree. But in this third generation the crossed plants of the same stock were in height to the self-fertilised only as 100 to 99, that is, they were practically equal. Nevertheless, when the eight crossed and eight self-fertilised plants were cut down and weighed, the former were to the latter in weight as 100 to 49! There can therefore be not the least doubt that the crossed plants of this species are greatly superior in vigour and luxuriance to the self-fertilised.

DIGITALIS

Digitalis purpurea. Watercolor on vellum by Lady Frances Howard,
A Catalogue of English Plants.

CROSS AND SELF-FERTILIZATION

Foxglove, with many species and varieties grown as biennials and perennials, is a lovely garden plant that is also well known for its toxicity and medicinal uses. Eighteenth century British physician and botanist William Withering is credited with being the first to systematically investigate the medicinal uses of *Digitalis*. His priority was disputed at the time by Erasmus Darwin, who also published on the use of this plant to treat certain diseases.[52] The plant yields digitoxin and other cardiac glycosides used today in the treatment of irregular heartbeat and related ailments.

The most common foxglove species, *Digitalis purpurea*, is a biennial native to much of Europe, cultivated in gardens everywhere and often escaping to grow in profusion. Its lovely deep-throated purple spotted flowers, arranged in a spiral up the stem, are mainly visited by bumblebees (known to Darwin and contemporaries as humble-bees). Cross-pollination of the many flowers is assured by protandry, where the flowers start out functionally male and then pass into a female stage. New flowers are continually produced at the top of the stem, creating a flower age (and sex) gradient from pollen-shedding male flowers at the top of the stem to stigma-receptive female flowers toward the bottom.

Darwin noticed that bumblebees visiting *Digitalis* started at the bottom-most flowers and worked their way up, collecting pollen at the top before leaving the plant and making a beeline to the female flowers at the bottom of another spike. In one of his memorable field studies, while on a family holiday in coastal Barmouth, Wales, he set out to investigate foxglove self- and cross-pollination. In a field where the plants abounded, he placed a net over some individuals to keep insects away while leaving others uncovered. He then hand-pollinated some flowers under the net and left others alone to see if they might self-pollinate—but to make the selfing test more realistic, he simulated the windy seaside conditions by violently shaking the plants, surely a curious spectacle to any passersby. He found that the netted plants produced far fewer seeds than those out in the open, even when hand-pollinated, once again confirming for Darwin the central importance of outcrossing.[53]

> **From: *The Effects of Cross and Self Fertilisation in the Vegetable Kingdom* (2nd ed., 1878)**

Digitalis purpurea. The flowers of the common Foxglove are proterandrous; that is, the pollen is mature and mostly shed before the stigma of the same flower is ready for fertilisation. This is effected by the larger humble-bees, which, whilst in search of nectar, carry pollen from flower to flower. The two upper and longer stamens shed their pollen before the two lower and shorter ones. The meaning of this fact probably is that the anthers of the longer stamens stand near to the stigma, so that they would be the most likely to fertilise it; and as it is an advantage to avoid self-fertilisation, they shed their pollen first, thus lessening the chance. There is, however, but little danger of self-fertilisation until the bifid stigma opens; for Hildebrand found that pollen placed on the stigma before it had opened produced no effect. The anthers, which are large, stand at first transversely with respect to the tubular corolla, and if they were to dehisce in this position they would smear with pollen the whole back and sides of an entering humble-bee in a useless manner; but the anthers twist round and place themselves longitudinally before they dehisce. The lower and inner side of the mouth of the corolla is thickly clothed with hairs, and these collect so much of the fallen pollen that I have seen the under surface of a humble-bee thickly dusted with it; but this can never be applied to the stigma, as the bees in retreating do not turn their under surfaces upwards. ...

I covered a plant growing in its native soil in North Wales with a net and fertilised six flowers each with its own pollen, and six others with pollen from a distinct plant growing within the distance of a few feet. The covered plant was occasionally shaken with violence, so as to imitate the effects of a gale of wind, and thus to facilitate as far as possible self-fertilisation. It bore ninety-two flowers and of these only twenty-four produced capsules; whereas almost all the flowers on the surrounding uncovered plants were fruitful. Of the twenty-four spontaneously self-fertilised capsules, only two contained their full complement of seed; six contained a moderate supply; and the remaining sixteen extremely few seeds. A little pollen adhering to the anthers after they had dehisced, and accidentally falling on the stigma when mature, must have been the means by which the above twenty-four flowers were partially self-fertilised; for the margins of the corolla in withering do not curl inwards, nor do the flowers in dropping off turn round on their axes, so as to bring the pollen-covered hairs, with which the lower surface is clothed, into contact with the stigma—by either of which means self-fertilisation might be effected.

Seeds from the above crossed and self-fertilised capsules, after germinating on bare sand, were planted in pairs on the opposite sides of five moderately-sized pots, which were kept in the greenhouse. The plants after a time appeared starved, and were therefore, without being disturbed, turned out of their pots, and planted in the open ground in two close parallel rows. ...

Seedlings raised from intercrossed flowers on the same plant, and others from flowers fertilised with their own pollen, were grown in the usual manner in competition with one another on the opposite sides of ten pots. ... In eight pots, in which the plants did not grow much crowded, the flower-stems on sixteen intercrossed plants were in height to those on sixteen self-fertilised plants, as 100 to 94. In the two other pots in which the plants grew much crowded, the flower-stems on nine intercrossed plants were in height to those on nine self-fertilised plants, as 100 to 90. That the intercrossed plants in these two latter pots had a real advantage over their self-fertilised opponents was well shown by their relative weights when cut down, which was as 100 to 78. The mean height of the flower-stems on the twenty-five intercrossed plants in the ten pots taken together was to that of the flower-stems on the twenty-five self-fertilised plants, as 100 to 92. Thus the intercrossed plants were certainly superior to the self-fertilised in some degree; but their superiority was small compared with that of the offspring from a cross between distinct plants over the self-fertilised, this being in the ratio of 100 to 70 in height. Nor does this latter ratio show at all fairly the great superiority of the plants derived from a cross between distinct individuals over the self-fertilised, as the former produced more than twice as many flower-stems as the latter and were much less liable to premature death.

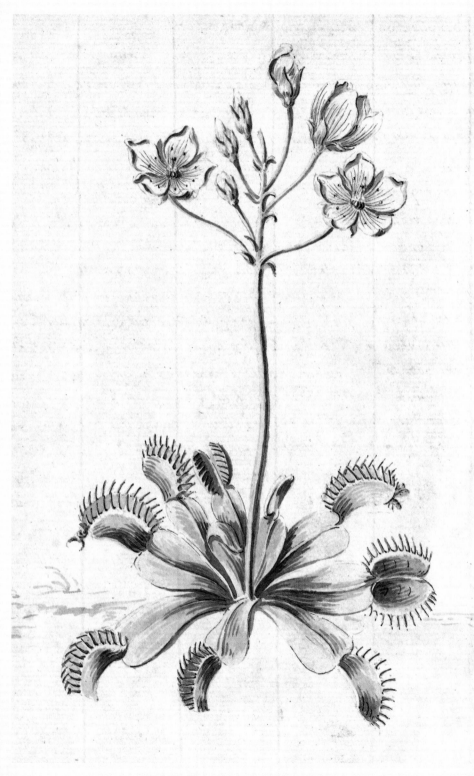

Dionaea muscipula. Pen-and-ink drawing by John Ellis, on letter to Carl Linnaeus.

Dionaea muscipula
VENUS FLY TRAP

....

DROSERACEAE—SUNDEW FAMILY

INSECTIVOROUS PLANTS

Venus fly trap is a relative of Darwin's beloved sundews (see p. 129) that grows in coastal wetlands of the Carolinas in the United States. First brought to the attention of naturalists in 1759 by North Carolina colonial governor Arthur Dobbs, this species has long been an object of fascination for botanists and non-botanists alike, owing to its remarkable snap-trap leaves. When Carl Linnaeus, father of modern taxonomy, first heard about this plant, he reportedly exclaimed *"miraculum naturae!"* and American naturalist William Bartram, who had made some of the earliest collections, declared it an "astonishing production!"[54]

Triggered by unwary insects, the paired lobes of the leaves snap shut in a flash—some 100 milliseconds, one of the fastest recorded movements in the plant kingdom. With rows of spiky processes along the outer margins of the leaf lobes, the leaves look for all the world like a botanical version of a toothed steel trap, misleading early observers into thinking that depredating insects were pierced or crushed to death when the lobes snapped shut. In fact, those spikes inter-digitate, rather than pierce, making a more effective trap for holding insects—but just why they are held was unknown to early naturalists and it took another century to figure out, something Darwin played a role in.

Despite its deadly attributes, eighteenth century naturalist Daniel Solander, who accompanied Joseph Banks on Captain James Cook's first voyage to the South Seas, thought that it "well deserved one of the names of the Goddess of Beauty," dubbing it *Dionaea* "from the beautiful Appearance of its Milk-white flowers, and the Elegance of its Leaves," according to London naturalist John Ellis in 1768. It was Ellis who added the specific epithet *muscipula*, the Latin word for mousetrap or flytrap.

Darwin's interest in Venus fly trap was two-fold. In the 1860s, he was riding his new "hobby horse" of plant movement hard, sure that the animal-like attribute of plant "irritability"—touch-sensitivity and movement—was more than analogical, pointing to an essential connection between plants and animals. To Darwin, the connection was evolutionary, and the fly trap's rapid movement (echoing the reflexive action of animals) and their carnivorous habit confirmed this view. One could hardly find a better plant-animal link than a green plant with a nervous system and stomach!*

He longed to study fly traps and compare them with sundews, but fly traps were still something of a rarity in Britain, and he was hard pressed to get his hands on specimens. Eventually, cajoling letters to Daniel Oliver, Keeper of the Herbarium at the Royal Botanic Gardens, Kew yielded a few traps and whole plants.[55] Asa Gray put him in touch with William Marriot Canby, in Wilmington, Delaware, and Mary Treat in Vineland, New Jersey, both of whom obligingly sent Darwin their own observations along with leaves holding trapped insects. Darwin eventually performed over a dozen digestion experiments, feeding various bits of albumen, gelatin, roasted meat, and cheese to determine the power of the leaves to dissolve the matter with the aid of their enzymatic secretions. He also localized the specialized hairs that trigger the traps, and teamed up with plant

* Modern evolutionary biologists agree that plants and animals—indeed, all species—share a common ancestor in deep time, but they would not agree with Darwin's assumptions about the animal-like characteristics of these plants. For example, their movements and touch-sensitivity are not based on a nervous system.

physiologist John Burdon-Sanderson, who devised a modified galvanometer to measure the electrical potential of the traps as they snap shut. Darwin was deeply impressed by this most animal-like plant—it's no surprise that he considered *Dionaea* "one of the most wonderful in the world."[56]

[From: *Insectivorous Plants* (1875)]

This plant, commonly called Venus' fly-trap, from the rapidity and force of its movements, is one of the most wonderful in the world. It is a member of the small family of the Droseraceae and is found only in the eastern part of North Carolina, growing in damp situations. The roots are small; those of a moderately fine plant which I examined consisted of two branches about 1 inch in length, springing from a bulbous enlargement. They probably serve, as in the case of *Drosera*, solely for the absorption of water; for a gardener who has been very successful in the cultivation of this plant

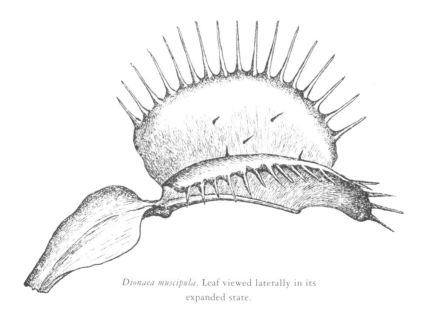

Dionaea muscipula. Leaf viewed laterally in its expanded state.

grows it like an epiphytic orchid, in well-drained damp moss without any soil. The form of the bilobed leaf, with its foliaceous footstalk, is shown in the accompanying drawing.

The two lobes stand at rather less than a right angle to each other. Three minute pointed processes or filaments, placed triangularly, project from the upper surfaces of both; but I have seen two leaves with four filaments on each side, and another with only two. These filaments are remarkable from their extreme sensitiveness to a touch, as shown not by their own movement, but by that of the lobes. The margins of the leaf are prolonged into sharp rigid projections which I will call spikes, into each of which a bundle of spiral vessels enters. The spikes stand in such a position that, when the lobes close, they inter-lock like the teeth of a rat-trap. ...

The sensitive filaments are formed of several rows of elongated cells, filled with purplish fluid. They are a little above the $\frac{1}{20}$ of an inch in length; are thin and delicate, and taper to a point. Towards the base there is constriction, formed of broader cells, beneath which there is an articulation, supported on an enlarged base, consisting of differently shaped polygonal cells. As the filaments project at right angles to the surface of the leaf, they would have been liable to be broken whenever the lobes closed together, had it not been for the articulation which allows them to bend flat down.

These filaments, from their tips to their bases, are exquisitely sensitive to a momentary touch. It is scarcely possible to touch them ever so lightly or quickly with any hard object without causing the lobes to close. A piece of very delicate human hair, $2\frac{1}{2}$ inches in length, held dangling over a filament, and swayed to and fro so as to touch it, did not excite any movement. But when a rather thick cotton thread of the same length was similarly swayed, the lobes closed. The sensitive filaments of *Dionaea* are not viscid, and the capture of insects can be assured only by their sensitiveness to a momentary touch, followed by the rapid closure of the lobes. ...

The upper surface of the lobes is thickly covered with small purplish, almost sessile glands. These have the power both of secretion and absorption; but unlike those of *Drosera*, they do not secrete until excited by the absorption of nitrogenous matter. ...

We will now consider the action of the leaves when insects happen to touch one of the sensitive filaments. This often occurred in my greenhouse, but I do not know whether insects are attracted in any special way by the leaves. They are caught in large numbers by the plant in its native country. As soon as a filament is touched, both lobes close with astonishing quickness; and as they stand at less than a right angle to each other, they have a good chance of catching any intruder. The angle between the blade and footstalk does not change when the lobes close. The chief seat of movement is near the midrib, but is not confined to this part; for, as the lobes come together, each curves inwards across its whole breadth; the marginal spikes however, not becoming curved. This movement of the whole lobe was well seen in a leaf to which a large fly had been given, and from which a large portion had been cut off the end of one lobe; so that the opposite lobe, meeting with no resistance in this part, went on curving inwards much beyond the medial line.

From the curving inwards of the two lobes, as they move towards each other, the straight marginal spikes intercross by their tips at first, and ultimately by their bases. The leaf is then completely shut and encloses a shallow cavity. If it has been made to shut merely by one of the sensitive filaments having been touched, or if it includes an object not yielding soluble nitrogenous matter, the two lobes retain their inwardly concave form until they re-expand. The re-expansion under these circumstances—that is when no organic matter is enclosed—was observed in ten cases. In all of these, the leaves re-expanded to about two-thirds of the full extent in 24 hrs. from the time of closure. Even the leaf from which a portion of one lobe had been cut off opened to a slight degree within this same time. How many times a leaf is capable of shutting and opening if no animal matter is left enclosed, I do not know; but one leaf was made to close four times, reopening afterwards, within six days, On the last occasion it caught a fly, and then remained closed for many days.

This power of reopening quickly after the filaments have been accidentally touched by blades of grass, or by objects blown on the leaf by the wind, as occasionally happens in its native place, must be of some importance to the plant; for as long as a leaf remains closed, it cannot of course capture an insect.

Dr. Canby, who observed in the United States a large number of plants which, although not in their native site, were probably more vigorous than my plants, informs me that he has "several times known vigorous leaves to devour their prey several times; but ordinarily twice, or, quite often, once was enough to render them unserviceable." Mrs. Treat, who cultivated many plants in New Jersey, also informs me that "several leaves caught successively three insects each, but most of them were not able to digest the third fly, but died in the attempt. Five leaves, however, digested each three flies, and closed over the fourth, but died soon after the fourth capture. Many leaves did not digest even one large insect." It thus appears that the power of digestion is somewhat limited, and it is certain that leaves always remain clasped for many days over an insect, and do not recover their power of closing again for many subsequent days. In this respect *Dionaea* differs from *Drosera*, which catches and digests many insects after shorter intervals of time.

We are now prepared to understand the use of the marginal spikes, which form so conspicuous a feature in the appearance of the plant and which at first seemed to me in my ignorance useless appendages. From the inward curvature of the lobes as they approach each other, the tips of the marginal spikes first intercross, and ultimately their bases. Until the edges of the lobes come into contact, elongated spaces between the spikes, varying from the $\frac{1}{15}$ to the $\frac{1}{10}$ of an inch (1.693 to 2.54 mm.) in breadth, according to the size of the leaf, are left open. Thus an insect, if its body is not thicker than these measurements, can easily escape between the crossed spikes, when disturbed by the closing lobes and in-creasing darkness; and one of my sons actually saw a small insect thus escaping. A moderately large insect, on the other hand, if it tries to escape between the bars will surely be pushed back again into its horrid prison with closing walls, for the spikes continue to cross more and more until the edges of the lobes come into contact. A very strong insect, however, would be able to free itself, and Mrs. Treat saw this effected by a rose-chafer (*Macrodactylus subspinosus*) in the United States. Now it would manifestly be a great disadvantage to the plant to waste many days in remaining clasped over a minute insect, and several additional days or weeks in afterwards recovering its sensibility; inasmuch as a minute insect would afford but little nutriment. It would be far better for the plant to wait for a time until a moderately large insect was captured, and to allow all the little ones to escape; and this advantage is secured by the slowly intercrossing marginal spikes, which act like the large meshes of a fishing-net, allowing the small and useless fry to escape.

As I was anxious to know whether this view was correct—and as it seems a good illustration of how cautious we ought to be in assuming, as I had done with respect to the marginal spikes, that any fully developed structure is useless—I applied to Dr. Canby. He visited the native site of the plant, early in the season, before the leaves had grown to their full size, and sent me fourteen leaves, containing naturally captured insects. Four of these had caught rather small insects, viz. three of them ants, and the fourth a rather small fly, but the other ten had all caught large insects, namely, five elaters, two chrysomelas, a curculio, a thick and broad spider, and a scolopendra. Out of these ten insects, no less than eight were beetles, and out of the whole fourteen there was only one, viz. a dipterous insect, which could readily take flight. *Drosera*, on the other hand, lives chiefly on insects which are good flyers, especially Diptera, caught by the aid of its viscid secretion. But what most concerns us is the size of the ten larger insects. Their average length from head to tail was .256 of an inch, the lobes of the leaves being on an average .53 of an inch in length, so that the insects were very nearly half as long as the leaves within which they were enclosed. Only a few of these leaves, therefore, had wasted their powers by capturing small prey, though it is probable that many small insects had crawled over them and been caught, but had then escaped through the bars.

Drosera rotundifolia. Watercolor by Lady Francis Howard,
A Catalogue of English Plants.

INSECTIVOROUS PLANTS

The Darwin family arrived in Bournemouth, southern England, in September 1862 for a seaside holiday to help twelve-year-old Lenny and his mother Emma convalesce from recent illnesses. Darwin was not happy, fretting over their recovery while wishing he had his work to distract him. Never one to sit still for long, he took the other kids on country rambles, but he wasn't impressed with what he saw. "This is a nice, but most barren country and I can find nothing to look at," he complained to Joseph Hooker. "Even the brooks and ponds produce nothing—The country is like Patagonia."[57] But as is true almost every-where, there was plenty to see once you looked closely, and he soon came upon a patch of round-leaved sundews, a curious plant with glistening dew-drop-covered leaves that he had begun to experiment with while vacationing in nearby Sussex two years earlier but then dropped as he became busy with other projects.

He immediately collected some to restart his sundew "feeding" experiments, beginning with bits of hair plucked from his own head and his toenails (both of which the picky plants decidedly rejected). It would have been interesting to have been a fly on the wall (at a safe distance from the sundews) in the Darwin's seaside cottage that day—picture the eminent naturalist carefully feeding his toenails to a plant! But as ever, there was method to the seeming madness. Darwin's revived interest in sundews snowballed into a series of ground-breaking investigations into plant carnivory, culminating thirteen years later in *Insectivorous Plants* (1875). Yet his motivation was not so much novel plant physiology, as interesting as that was, but more the way these plants emulated animals—an emulation that spoke of ancestral evolutionary union of plants and animals. Emma Darwin perhaps expressed it best when she remarked on her husband's sundew fascination to Mary Lyell, wife of the geologist Charles Lyell, "I suppose he hopes to end in proving it to be an animal."[58] But more than that, he was sure these plants taught lessons in evolutionary gradualism to boot. What did the simple movement and sensory abilities of *Drosera* say about the evolutionary steps leading to their far more animal-like relatives, Venus fly trap (see p. 121)? "I began this work on *Drosera* in relation to gradation as throwing light on *Dionaea*," he admitted to Asa Gray.[59]

Drosera rotundifolia, common in sunny boggy areas throughout the northern hemisphere, is one of nearly 200 species in the genus worldwide, now divided into about a dozen sub-genera. All the species have glandular hairs that glisten like dew in the sun and enable them to capture and digest insects, a vital source of nitrogen in acidic, nutrient-poor soil. At the time of Darwin's experiments, it was well known that insects got trapped on the leaves by the sticky mucilage, but no one knew if they were digested—some thought the leaves acted as botanical flypaper, protecting the plant. Darwin was able to prove that the insects do indeed become food—*Drosera* is acknowledged as the first plant genus in which carnivory was confirmed—but more than that, he and his son Francis later demonstrated just how beneficial carnivory is to sundews, showing in a controlled experiment that plants "fed" insects grew more vigorously and produced more flowers and seeds than "starved" plants shielded from insects by netting. Francis Darwin published this study in 1880 in the *Journal of the Linnean Society* and in the second edition of *Insectivorous Plants* (revised by Francis in 1888).

In his first series of experiments in the 1860s, Darwin sought to determine the nature of the dewy secretion acidic enough to break down insects. He found it was similar to the digestive pepsin and acids in animals, and that it also has antiseptic properties, preventing growth of mold. For his experiments, Darwin filled his greenhouse with sundews, raiding his kitchen and medicine cabinet for all sorts of things to try out on them as food. He tested with nitrogenous and non-nitrogenous substances, liquids and solids—albumen, milk, olive oil, boiled peas, ammonia, hydrochloric acid, glycerin, turpentine, quinine, cobra venom, and more. In his usual fashion, he encouraged others to try experiments too. Writing to Daniel Oliver at Kew Gardens, Darwin marveled how his sundews could somehow detect nitrogen in fluids. Noting that "Our *Drosera* likes milk better than any other drink,"[60] he suggested that Oliver try feeding droplets of milk and saliva to an Australian species for comparison. Ultimately, his investigations turned into a characteristically collaborative affair: another half dozen *Drosera* species were analyzed with the help of Oliver and others, clergyman Henry M. Wilkinson sent observations of trapped insects, John Burdon-Sanderson, professor of physiology at University College London, studied their electrical impulses, and Darwin's sons George and Francis helped with experiments and drew illustrations of the leaves in various stages of entrapping prey with their dew-tipped glandular hairs—referred to by Darwin as "tentacles," a reflection of his tendency to see the animal in these plants. Ultimately, he dedicated eleven of the eighteen chapters of *Insectivorous Plants* to *Drosera*, mainly his beloved *D. rotundifolia*: "a wonderful plant, or rather a most sagacious animal."[61]

[From: *Insectivorous Plants* (2nd ed., 1888)]

During the summer of 1860, I was surprised by finding how large a number of insects were caught by the leaves of the common sun-dew (*Drosera rotundifolia*) on a heath in Sussex. I had heard that insects were thus caught, but knew nothing further on the subject. I gathered by chance a dozen plants, bearing fifty-six fully expanded leaves, and on thirty-one of these, dead insects or remnants of them adhered; and, no doubt, many more would have been caught afterwards by these same leaves, and still more by those as yet not expanded. On one plant, all six leaves had caught their prey; and on several plants, very many leaves had caught more than a single insect. On one large leaf, I found the remains of thirteen distinct insects. Flies (Diptera) are captured much oftener than other insects. The largest kind which I have seen caught was a small butterfly (*Caenonympha pamphilus*); but the Rev. H.M. Wilkinson informs me that he found a large living dragon-fly with its body firmly held by two leaves. As this plant is extremely common in some districts, the number of insects thus annually slaughtered must be prodigious. Many plants cause the death of insects, for instance the sticky buds of the horse-chestnut (*Aesculus hippocastanum*), without thereby receiving, as far as we can perceive, any advantage; but it was soon evident that *Drosera* was excellently adapted for the special purpose of catching insects, so that the subject seemed well worthy of investigation. ...

It is necessary, in the first place, to describe briefly the plant. It bears from two or three to five or six leaves, generally extended more or less horizontally but sometimes standing vertically upwards. The shape and general appearance of a leaf is shown, as seen from above... The whole upper surface is covered with gland-bearing filaments, or tentacles, as I shall call them, from their manner of acting. ... The glands are each surrounded by large drops of extremely viscid secretion, which, glittering in the sun, have given rise to the plant's poetical name of the sun-dew.

The tentacles on the central part of the leaf or disc are short and stand upright, and their pedicels are green. Towards the margin, they become longer and longer and more inclined outwards, with their pedicels of a purple colour. Those on the extreme margin project in the same plane with the leaf, or more commonly are considerably reflexed. A few tentacles spring from the base of the footstalk or petiole, and these are the longest of all, being sometimes nearly ¼ of an inch in length. ...

[left] Leaf viewed from above, enlarged four times.
[center] Leaf with all the tentacles closely inflected,
from immersion in a solution of phosphate of ammonia.
[right] Leaf with the tentacles on one side inflected
over a bit of meat placed on the disc.

[Since the publication of the first edition, several experiments have been made to determine whether insectivorous plants are able to profit by an animal diet. ...

... My [Francis Darwin's] experiments were begun in June 1877, when the plants were collected and planted in six ordinary soup-plates. Each plate was divided by a low partition into two sets, and the least flourishing half of each culture was selected to be "fed," while the rest of the plants were destined to be "starved." The plants were prevented from catching insects for themselves by means of a covering of fine gauze, so that the only animal food which they obtained was supplied in very minute pieces of roast meat given to the "fed" plants but withheld from the "starved" ones. After only ten days the difference between the fed and starved plants was clearly visible: the fed plants were of brighter green and the tentacles of a more lively red. At the end of August the plants were compared by number, weight, and measurement. ...

[The results] show clearly enough that insectivorous plants derive great advantage from animal food. It is of interest to note that the most striking difference between the two sets of plants is seen in what relates to reproduction—i.e. in the flower-stems, the capsules, and the seeds. ...

... Both starved and fed plants were kept without food until April 3rd, when it was found that the average weights per plant were 100 for the starved, 213.0 for the fed. This proves that the fed plants had laid by a far greater store of reserve material in spite of having produced nearly four times as much seed. ...—F. D.]

... It will be sufficient here to recapitulate, as briefly as I can, the chief points. In the first chapter, a preliminary sketch was given of the structure of the leaves and of the manner in which they capture insects. This is effected by drops of extremely viscid fluid surrounding the glands and by the inward movement of the tentacles. As the plants gain most of their nutriment by this means, their roots are very poorly developed; and they often grow in places where hardly any other plant except mosses can exist. The glands have the power of absorption, besides that of secretion. They are extremely sensitive to various stimulants, namely repeated touches, the pressure of minute particles, the absorption of animal matter and of various fluids, heat, and galvanic action. A tentacle with a bit of raw meat on the gland has been seen to begin bending in 10 s., to be strongly incurved in 5 m., and to reach the centre of the leaf in half an hour. The blade of the leaf often becomes so much inflected that it forms a cup, enclosing any object placed on it.

A gland, when excited, not only sends some influence down its own tentacle, causing it to bend, but likewise to the surrounding tentacles, which become incurved; so that the bending place can be acted on by an impulse received from opposite directions, namely from the gland on the summit of the same tentacle and from one or more glands of the neighbouring tentacles. Tentacles, when inflected, re-expand after a time, and during this process the glands secrete less copiously, or become dry. As soon as they begin to secrete again, the tentacles are ready to re-act; and this may be repeated at least three, probably many more times. ...

Movement ensues if a gland is momentarily touched three or four times; but if touched only once or twice, though with considerable force and with a hard object, the tentacle does not bend. The plant is thus saved from much useless movement, as during a high wind the glands can hardly escape being occasionally brushed by the leaves of surrounding plants. Though insensible to a single touch, they are exquisitely sensitive, as just stated, to the slightest pressure if prolonged for a few seconds; and this capacity is manifestly of service to the plant in capturing small insects. Even gnats, if they rest on the glands with their delicate feet, are quickly and securely embraced. The glands are insensible to the weight and repeated blows of drops of heavy rain, and the plants are thus likewise saved from much useless movement. ...

In the fifth chapter, the results of placing drops of various nitrogenous and non-nitrogenous organic fluids on the discs of leaves were given, and it was shown that they detect with almost unerring certainty the presence of nitrogen. These results led me to inquire whether *Drosera* possessed the power of dissolving solid animal matter. The experiments proving that the leaves are capable of true digestion and that the glands absorb the digested matter, are given in detail in the sixth chapter. These are, perhaps, the most interesting of all my observations on *Drosera*, as no such power was before distinctly known to exist in the vegetable kingdom. It is likewise an interesting fact that the glands of the disc, when irritated, should transmit some influence to the glands of the exterior tentacles, causing them to secrete more copiously and the secretion to become acid, as if they had been directly excited by an object placed on them. The gastric juice of animals contains, as is well known, an acid and a ferment, both of which are indispensable for digestion, and so it is with the secretion of *Drosera*. When the stomach of an animal is mechanically irritated, it secretes an acid, and when particles of glass or other such objects were placed on the glands of *Drosera*, the secretion, and that of the surrounding and untouched glands, was increased in quantity and became acid. But the stomach of an animal does not secrete its proper ferment, pepsin, until certain substances, called peptogenes, are absorbed; and it appears from my experiments that some matter must be absorbed by the glands of *Drosera* before they secrete their proper ferment. That the secretion does contain a ferment which acts only in the presence of an acid on solid animal matter, was clearly proved by adding minute doses of an alkali, which entirely arrested the process of digestion, this immediately recommencing as soon as the alkali was neutralised by a little weak hydrochloric acid. From trials made with a large number of substances, it was found that those which the secretion of *Drosera* dissolves completely, or partially, or not at all, are acted on in exactly the same manner by gastric juice. We may, therefore, conclude that the ferment of *Drosera* is closely analogous to, or identical with, the pepsin of animals. ...

Most of the acids which were tried, though much diluted (one part to 437 of water) and given in small doses, acted powerfully on *Drosera*; nineteen, out of the twenty-four, causing the tentacles to be more or less inflected. Most of them, even the organic acids, are poisonous, often highly so; and this is remarkable, as the juices of so many plants contain acids. ... Many acids excite the glands to secrete an extraordinary quantity of mucus; and the protoplasm within their cells seems to be often killed, as may be inferred from the surrounding fluid soon becoming pink.

In the ninth chapter, the effects of the absorption of various alkaloids and certain other substances were described. Although some of these are poisonous, yet as several, which act powerfully on the nervous system of animals, produce no effect on *Drosera*; we may infer that the extreme sensibility of the glands, and their power of transmitting an influence to other parts of the leaf, causing movement, or modified secretion, or aggregation, does not depend on the presence of a diffused element, allied to nerve-tissue. One of the most remarkable facts is that long immersion in the poison of the cobra-snake does not in the least check, but rather stimulates, the spontaneous movements of the protoplasm in the cells of the tentacles. Solutions of various salts and acids behave very differently in delaying or in quite arresting the subsequent action of a solution of phosphate of ammonia. Camphor dissolved in water acts as a stimulant, as do small doses of certain essential oils, for they cause rapid and strong inflection. Alcohol is not a stimulant. The vapours of camphor, alcohol, chloroform, sulphuric and nitric ether, are poisonous in moderately large doses, but in small doses serve as narcotics or anaesthetics, greatly delaying the subsequent action of meat. But some of these vapours also act as stimulants, exciting rapid, almost spasmodic movements in the tentacles. ...

In the tenth chapter, it was shown that the sensitiveness of the leaves appears to be wholly confined to the glands and to the immediately underlying cells. It was further shown that the motor impulse and other forces or influences proceeding from the glands when excited pass through the cellular tissue and not along the fibro-vascular bundles. A gland sends its motor impulse with great rapidity down the pedicel of the same tentacle to the basal part which alone bends. The impulse, then passing onwards, spreads on all sides to the surrounding tentacles, first affecting those which stand nearest and then those farther off. But by being thus spread out, and from the cells of the disc not being so much elongated as those of the tentacles, it loses force, and here travels much more slowly than down the pedicels. Owing also to the direction and form of the cells, it passes with greater ease and celerity in a longitudinal than in a transverse line across the disc. ...

I have now given a brief recapitulation of the chief points observed by me, with respect to the structure, movements, constitution, and habits of *Drosera rotundifolia*; and we see how little has been made out in comparison with what remains unexplained and unknown.

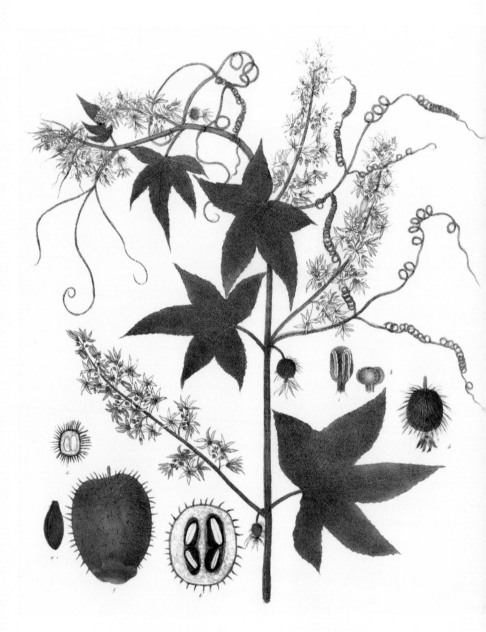

Echinocystis lobata. Lithograph by George Endicott, printed in
Natural History of New York.

Echinocystis
BUR CUCUMBER

———— ...... ————

CUCURBITACEAE—
GOURD AND SQUASH FAMILY

CLIMBING PLANTS

Echinocystis lobata is an annual vine with delightfully fragrant flowers found throughout most of North America. It bears several common names, including wild cucumber and balsam apple, while its Latin name is derived from the Greek words *echinos* ("hedgehog") and *cystis* ("bladder"), describing the small prickle-covered fruits. This sprawling plant drapes its broad, lobed leaves and spiked clusters of star-like flowers while relying on its tendrils to climb over shrubs.

Bur cucumber led to Darwin's research on climbing plants. He had become curious about climbers after reading an article by Harvard botanist Asa Gray, reporting experiments on the touch-sensitivity of the tendrils of the related vine *Sicyos angulatus*.*[62] Encouraging his friend to study the climbers, Gray sent Darwin *Echinocystis* seeds while joking that he doubted there was "warmth and sunshine enough in England to a get a sensible movement."[63] But Darwin had some germination success, and it wasn't long before he was able to observe the remarkable tendril movements, inspiring him to study well over 100 species of climbing plants in the next several years, more than half of them tendril-bearers.

It was by carefully watching *Echinocystis* that Darwin first noticed the gyrating motions he was to later dub *circumnutation*. "This may be

---

* Gray was investigating a report of touch-sensitivity in tendrils by German botanist Hugo von Mohl (Mohl 1827).

common phenomenon for what I know," he wrote to his friend Joseph Hooker at Kew, "but it confounded me quite when I began to observe the irritability [sensitivity] of the tendrils. ... The result is pretty, for the plant every 1½ or 2 hours sweeps a circle (according to length of bending shoot and length of tendril) of from 1 foot to 20 inches in diameter, and immediately that the tendril touches any object its sensitiveness causes it immediately to seize it."[64]

This incessant movement of the tendrils is behind the plant's uncanny ability to "seek" and find a support to climb, giving the impression of an almost animal awareness and intentionality. A local gardener who helped Darwin with experiments certainly thought so—Darwin related to Hooker how "a clever gardener, my neighbour, who saw the plant on my table last night, said 'I believe, Sir, the tendrils can see, for wherever I put the plant, it finds out any stick near enough.' I believe the above is the explanation, viz that it sweeps slowly round and round."[65]

When *Echinocystis* tendrils coil and catch a support, they slowly become twisted in opposite directions, resulting in two oppositely spiraling sections approximately equal in length, separated by a short straight segment. Bur cucumber is not the only vine to do this; Darwin's son Francis illustrated the phenomenon in *Climbing Plants* with a Bryony tendril.

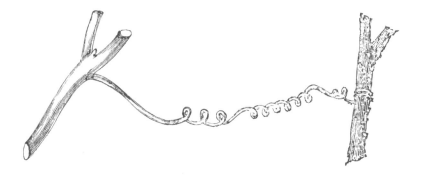

A caught tendril of *Bryonia dioica*, spirally contracted in reverse directions.

One effect of this reverse twisting is to pull the plant closer to the support, and another is to create a spring of the tendril, firmly anchored to the support yet allowing some give—presumably useful in windy conditions.

In keeping with Darwin's emphasis on animal-like qualities of plants, he used a human analogy in his discussion of how *Echinocystis* tendrils slowly but surely grasp a support: "It drags itself onwards by an insensibly slow, alternate movement, which may be compared to that of a strong man suspended by the ends of his fingers to a horizontal pole, who works his fingers onwards until he can grasp the pole with the palm of his hand."[66]

From: *The Movements and Habits of Climbing Plants* (2nd ed., 1875)

Echinocystis lobata.—Numerous observations were made on this plant (raised from seed sent me by Prof. Asa Gray), for the spontaneous revolving movements of the internodes and tendrils were first observed by me in this case, and greatly perplexed me. My observations may now be much condensed. I observed thirty-five revolutions of the internodes and tendrils; the slowest rate was 2 hrs., and the average rate, with no great fluctuations, 1 hr. 40 m. Sometimes I tied the internodes, so that the tendrils alone moved; at other times I cut off the tendrils whilst very young, so that the internodes revolved by themselves; but the rate was not thus affected. The course generally pursued was with the sun, but often in an opposite direction. Sometimes the movement during a short time would either stop or be reversed; and this apparently was due to interference from the light, as, for instance, when I placed a plant close to a window. In one instance, an old tendril, which had nearly ceased revolving, moved in one direction, whilst a young tendril above moved in an opposite course. The two uppermost internodes alone revolve; and as soon as the lower one grows old, only its upper part continues to move. The ellipses or circles swept by the summits of the internodes are about three inches in diameter; whilst those swept by the tips of the tendrils, are from 15 to 16 inches in diameter. During the revolving movement, the internodes become successively curved to all points of the compass; in one part of their course they are often inclined, together with the tendrils, at about 45° to the horizon, and in another part stand vertically up. There was something in the appearance of the revolving internodes which continually gave the false impression that their movement was due to the weight of the long and spontaneously revolving tendril; but, on cutting off the latter with sharp scissors, the top of the shoot rose only a little, and went on revolving. This false appearance is apparently due to the internodes and tendrils all curving and moving harmoniously together.

A revolving tendril, though inclined during the greater part of its course at an angle of about 45° (in one case of only 37°) above the horizon, stiffened and straightened itself from tip to base in a certain part of its course, thus becoming nearly or quite vertical. I witnessed this repeatedly; and it occurred both when the supporting internodes were free and when they were tied up; but was perhaps most conspicuous in the latter case, or when the whole shoot happened to be much inclined. The tendril forms a very acute angle with the projecting extremity of the stem or shoot; and the stiffening always occurred as the tendril approached and had to pass over the shoot in its circular course. If it had not possessed and exercised this curious power, it would infallibly have struck against the extremity of the shoot and been arrested. As soon as the tendril with its three branches begins to stiffen itself in this manner and to rise from an inclined into a vertical position, the revolving motion becomes more rapid; and as soon as the tendril has succeeded in passing over the extremity of the shoot or point of difficulty, its motion, coinciding with that from its weight, often causes it to fall into its previously inclined position so quickly that the apex could be seen travelling like the minute hand of a gigantic clock.

The tendrils are thin, from 7 to 9 inches in length, with a pair of short lateral branches rising not far from the base. The tip is slightly and permanently curved, so as to act to a limited extent as a hook. The concave side of the tip is highly sensitive to a touch; but not so the convex side. I repeatedly proved this difference by lightly rubbing four or five times the convex side of one tendril, and only once or twice the concave side of another tendril, and the latter alone curled inwards. In a few hours afterwards, when the tendrils which had been rubbed on the concave side had straightened themselves, I reversed the process of rubbing, and always with the same result. After touching the concave side, the tip becomes sensibly curved in one or two minutes; and subsequently, if the touch has been at all rough, it coils itself into a helix. But the helix will, after a time, straighten itself, and be again ready to act. ...

The revolving movement of a tendril is not stopped by the curving of its extremity after it has been touched. When one of the lateral branches has firmly clasped an object, the middle branch continues to revolve. When a stem is bent down and secured, so that the tendril depends but is left free to move, its previous revolving movement is nearly or quite stopped; but it soon begins to bend upwards, and as soon as it has become horizontal, the revolving movement recommences. I tried this four times; the tendril generally rose to a horizontal position in an hour or an hour and a half; but in one case, in which a tendril depended at an angle of 45° beneath the horizon, the uprising took two hours; in half an hour afterwards it rose to 23° above the horizon and then recommenced revolving. This upward movement is independent of the action of light, for it occurred twice in the dark, and on another occasion the light came in on one side alone. The movement no doubt is guided by opposition to the force of gravity, as in the case of the ascent of the plumules of germinating seeds.

A tendril does not long retain its revolving power; and as soon as this is lost, it bends downwards and contracts spirally. After the revolving movement has ceased, the tip still retains for a short time its sensitiveness to contact, but this can be of little or no use to the plant.

Epipactis latifolia. Watercolor by Elizabeth Wharton, *British Flowers*.

Epipactis
HELLEBORINE ORCHID

——— ...... ———

ORCHIDACEAE—ORCHID FAMILY

ORCHIDS, FORMS OF FLOWERS, POLLINATION

Epipactis is a large orchid genus of some seventy species, placed in the tribe Neotteae, a group with free-standing anthers. Darwin was curious about *Epipactis palustris* (the marsh helleborine), and in 1860, he requested fresh flowers from the botanist Alexander More from the Isle of Wight, so he could examine and manipulate the various parts of the flowers to determine their pollination mechanism. He also asked More to observe pollinators and do experiments on flowers in their native habitat to confirm his observations on the movement of the labellum and pollinia in promoting pollination. It was a poor time for pollination studies in the field, however, as the summer of 1860 was one of the coldest and wettest on record in the British Isles. "The only chance of seeing insects at work," he wrote More hopefully, "would be the first bright day after this miserable weather or a bright gleam of few hours in middle of one of our gloomy days."[67] Darwin later noted that "we see the injurious effects of the extraordinary cold and wet season of 1860 in the infrequency of the visits of insects."[68] His son William made observations on the Isle of Wight several years later, describing mainly honeybees pollinating the plants.

A related species, *Epipactis latifolia* (= *E. helleborine*) appeared in Darwin's yard unexpectedly, prompting him to report it to the *Gardeners' Chronicle*.[69] Although not rare then or now (it has become widely naturalized even in North America, where it is considered a weed by some), what was surprising, as he described in his report, was *where*

it appeared—right in the middle of his gravel walking path, now
known as the sand-walk. Given the history of disturbance, first as a
carriage road and then as his gravel footpath, had this orchid been
lying dormant for years? Or maybe a seed was blown there from some
far-flung locale?

Not one to waste an opportunity, he carefully observed the guest
orchid over several seasons, and found wasps to be its main pollinators, sucking nectar out of the big cup-shaped labellum while pollen
masses became attached to their foreheads to be carried to other
flowers.[70] He would have surely marveled at more modern studies
showing that, like many flowering plants, *E. helleborine* orchids have
additional tricks up their sleeves (er, flowers) to facilitate pollination.
They produce a chemical cocktail in their nectar that includes both
attractants and compounds with narcotic or soporific qualities that
could function to keep the wasps in the vicinity, increasing the likeli-
hood of visitation to additional flowers.[71]

[**From: *The Various Contrivances by Which Orchids
are Fertilised by Insects* (2nd ed., 1877)**]

Epipactis palustris. The flowers stand out (fig. A) almost horizontally from
the stem. The labellum is curiously shaped, as may be seen in the drawings: the distal half, which projects beyond the other petals and forms an
excellent landing-place for insects, is joined to the basal half by a narrow
hinge, and naturally is turned a little upwards, so that its edges pass within
the edges of the basal portion. So flexible and elastic is the hinge that the
weight of even a fly, as Mr. More informs me, depresses the distal portion; it is represented in fig. B in this state; but when the weight is removed
it instantly springs up to its former position (fig. A), and with its curious
medial ridges partly closes the entrance into the flower. The basal portion
of the labellum forms a cup, which at the proper time is filled with nectar.

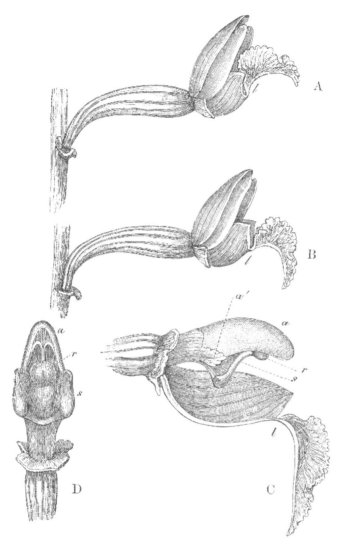

Epipactis palustris. A. Side view of flower (with the lower sepals alone removed) in its natural position. B. Side view of flower, with the distal portion of the labellum depressed, as if by the weight of an insect. C. Side view of flower, with all the sepals and petals removed, excepting the labellum, of which the near side has been cut away; the massive anther is seen to be of large size. D. Front view of column, with all the sepals and petals removed

a. anther, with the two open cells seen in the front view. *a'.* rudimentary anther, or auricle, referred to in a future chapter. *r.* rostellum. *s.* stigma. *l.* labellum.

Now let us see how all the parts, which I have been obliged to describe in detail, act. When I first examined these flowers, I was much perplexed: trying in the same manner as I should have done with a true *Orchis*, I slightly pushed the protuberant rostellum downwards, and it was easily ruptured; some of the viscid matter was withdrawn, but the pollinia remained in their cells. Reflecting on the structure of the flower, it occurred to me that an insect in entering one in order to suck the nectar would depress the distal portion of the labellum, and consequently would not touch the rostellum; but that, when within the flower, it would be almost compelled, from the springing up of this distal half of the labellum, to rise a little upwards and back out parallel to the stigma. I then brushed the rostellum lightly upwards and backwards with the end of a feather and other such objects; and it was pretty to see how easily the membranous cap of the rostellum came off, and how well from its elasticity it fitted any object, whatever its shape might be, and how firmly it clung to the object owing to the viscidity of its under surface. Large masses of pollen, adhering by the elastic threads to the cap of the rostellum were at the same time withdrawn.

Nevertheless, the pollen-masses were not removed nearly so cleanly as those which had been naturally removed by insects. I tried dozens of flowers, always with the same imperfect result. It then occurred to me that an insect in backing out of the flower would naturally push with some part of its body against the blunt and projecting upper end of the anther, which overhangs the stigmatic surface. Accordingly, I so held a brush that, whilst brushing upwards against the rostellum, I pushed against the blunt solid end of the anther (see fig. C); this at once eased the pollinia, and they were withdrawn in an entire state. At last I understood the mechanism of the flower. ...

Epipactis latifolia.—This species agrees with the last in most respects. The rostellum, however, projects considerably further beyond the face of the stigma, and the blunt upper end of the anther less so. The viscid matter lining the elastic cap of the rostellum takes a longer time to get dry. The upper petals and sepals are more widely expanded than in *E. palustris*: the distal portion of the labellum is smaller and is firmly united to the basal portion so that it is not flexible and elastic; it apparently serves only as a landing-place for insects. The fertilisation of this species depends simply on an insect striking in an upward and backward direction the highly-protuberant rostellum, which it would be apt to do when retreating from the flower after having sucked the copious nectar in the cup of the labellum. Apparently it is not at all necessary that the insect should push upwards the blunt upper end of the anther; at least I found that the pollinia could be removed easily by simply dragging off the cap of the rostellum in an upward or backward direction.

As some plants grew close to my house, I have been able to observe here and elsewhere their manner of fertilisation during several years. Although hive-bees and humble-bees of many kinds were constantly flying over the plants, I never saw a bee or any Dipterous insect visit the flowers; but in Germany, Sprengel caught a fly with the pollinia of this plant attached to its back. On the other hand, I have repeatedly observed the common wasp (*Vespa sylvestris*) sucking the nectar out of the open cup-shaped labellum. I thus saw the act of fertilisation effected by the pollen-masses being removed by the wasps and afterwards carried attached to their foreheads to other flowers. ... It is very remarkable that the sweet nectar of this *Epipactis* should not be attractive to any kind of bee. If wasps were to become extinct in any district, so probably would the *Epipactis latifolia*.

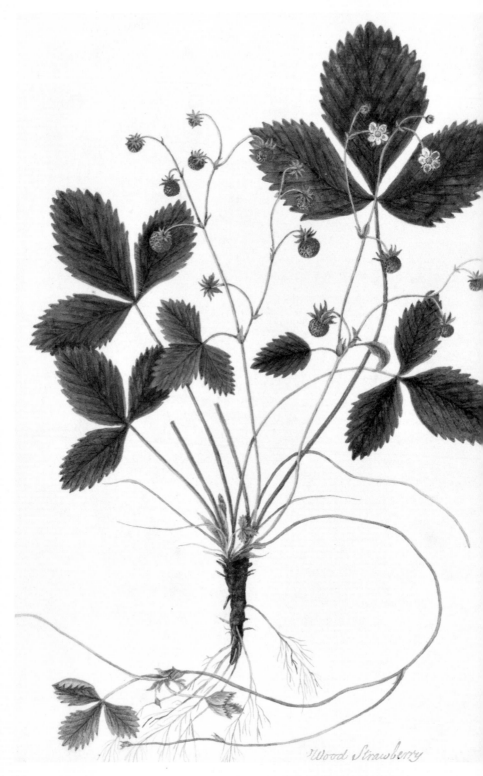

Wood Strawberry

Fragaria vesca. Watercolor by Elizabeth Blackwell, *A Curious Herbal.*

Fragaria

STRAWBERRY

——— ...... ———

ROSACEAE—ROSE FAMILY

VARIATION, FORMS OF FLOWERS, POLLINATION, PLANT MOVEMENT

The kitchen garden at Down House, Darwin's family home in Kent, England, always presented low-hanging fruit, so to speak, for Darwin's botanical investigations, and several species were studied from different biological angles. Strawberries, *Fragaria*, are a good example—hybridization, flower structure, and movement of the runners (stolons) during growth provided research topics at different times, all exemplifying for him distinct lines of evidence for evolution by natural selection.

Darwin argued that domestication, through artificial selection, was a compelling analog to the way natural selection operates—one reason that *On the Origin of Species* opens with a chapter on "Variation Under Domestication." The key word there is "variation," as heritable variation is raw material for selection, artificial or natural. Strawberries are not discussed in *Origin*, but in his 1868 book *The Variation of Animals and Plants Under Domestication*, Darwin invited readers to compare the large and plump garden strawberries then available with their diminutive wild relatives. What accounts for the difference? Improvement through crossing (which mixes things up and introduces more variation) combines with selection to yield ever-larger fruits over time.

Investigating strawberry species and varieties, and their propensity to hybridize, Darwin employed his favored crowd-sourcing method—publishing a letter asking for help. In the *Journal of Horticulture, Cottage Gardener, and Country Gentleman*, he wrote:

> *Will any of your correspondents who have attended to the history*
> *of the Strawberry kindly inform me whether any of the kinds now,*
> *or formerly, cultivated have been raised from a cross between any of*
> *the Woods or Alpines with the Scarlets, Pines, and Chilis? Also,*
> *whether any one has succeeded in getting any good from a cross*
> *between the Hautbois and any other kind? … I should feel greatly*
> *indebted to any one who would take the trouble to inform me on*
> *this head.*[72]

A related interest was the curious fact that strawberries vary in their flower structure, first noticed by French botanist Antoine Nicolas Duchesne, who cultivated a diversity of strawberry species and varieties in the Royal Botanical Gardens at Versailles. Some individuals are unisexual, bearing male-only or female-only flowers (dioecious, in modern terms), while others are hermaphroditic, with flowers bearing both stamens and pistils (monecious). In this, Darwin saw evolution in action, noting in *Forms of Flowers* (1877) of strawberries' "tendency to the separation of the sexes."

From: *The Variation of Animals and Plants Under Domestication, vol* 1. (1868)

Strawberries (Fragaria).— This fruit is remarkable on account of the number of species which have been cultivated, and from their rapid improvement within the last fifty or sixty years. Let any one compare the fruit of one of the largest varieties exhibited at our Shows with that of the wild wood strawberry, or, which will be a fairer comparison, with the somewhat larger fruit of the wild American Virginian Strawberry, and he will see what prodigies horticulture has effected. The number of varieties has likewise increased in a surprisingly rapid manner. Only three kinds were known in France, in 1746, where this fruit was early cultivated. In 1766, five species had been introduced, the same which are now cultivated, but only five

varieties of *Fragaria vesca*, with some sub-varieties, had been produced. At
the present day, the varieties of the several species are almost innumerable.

Much has been written on the sexes of strawberries; the true Hautbois
properly bears the male and female organs on separate plants and was
consequently named by Duchesne *dioica*; but it frequently produces hermaphrodites; and Lindley, by propagating such plants by runners, at the
same time destroying the males, soon raised a self-prolific stock. The other
species often show a tendency towards an imperfect separation of the
sexes, as I have noticed with plants forced in a hot-house. Several English
varieties, which in this country are free from any such tendency, when cultivated in rich soils under the climate of North America, commonly produce
plants with separate sexes. Thus a whole acre of Keen's Seedlings in the
United States has been observed to be almost sterile from the absence of
male flowers; but the more general rule is that the male plants overrun the
females. Some members of the Cincinnati Horticultural Society, especially
appointed to investigate this subject, report that "few varieties have the
flowers perfect in both sexual organs," &c. The most successful cultivators
in Ohio plant for every seven rows of "pistillata," or female plants, one row
of hermaphrodites, which afford pollen for both kinds; but the hermaphrodites, owing to their expenditure in the production of pollen, bear less
fruit than the female plants.

Darwin later turned his attention to the growth of strawberry stolons
or runners, horizontal stems that reach out to root beyond the main
stems, enabling the plant to spread vegetatively. He had been studying the circumnutation of aerial stems and wondered if stolons also
rotate as they grow. It turned out that they do. In *Movement of Plants*
(1880), Darwin described how strawberry stolons twist and rotate
as they maneuver around objects in their path. He documented this
movement by mounting a sheet of glass above or beside a plant and
applying black sealing wax to a fine glass filament attached to the
part of the plant he wanted to watch. He would then mark a black dot

on a white card fixed in position on a stick just beneath the needle. Noting the location of the dot on the glass plane, he would mark each movement over a period of time, usually over one or two days. He would then connect the dots to make a record of the movement.[73]

[**From: *The Power of Movement in Plants* (1880)**]

Stolons consist of much elongated, flexible branches, which run along the surface of the ground and form roots at a distance from the parent-plant. They are therefore of the same homological nature as stems. ...

Fragaria (cultivated garden var.): A plant growing in a pot had emitted a long stolon; this was supported by a stick, so that it projected for the length of several inches horizontally. A glass filament bearing two minute triangles of paper was affixed to the terminal bud, which was a little upturned; and its movements were traced during 21 h., as shown in [the left-hand figure]. In the course of the first 12 h., it moved twice up and twice down in some- what zigzag lines, and no doubt travelled in the same manner during the night. On the following morning after an interval of 20 h., the apex stood a little higher than it did at first, and this shows that the stolon had not been acted on within this time by geotropism; nor had its own weight caused it to bend downwards.

On the following morning (19th) the glass filament was detached and refixed close behind the bud, as it appeared possible that the circumnu- tation of the terminal bud and of the adjoining part of the stolon might be different. The movement was now traced during two consecutive days. ... During the first day, the filament travelled in the course of 14 h. 30 m. five times up and four times down, besides some lateral movement. On the 20th, the course was even more complicated, and can hardly be followed in the figure; but the filament moved in 16 h. at least five times up and five times down, with very little lateral deflection. The first and last dots made on this second day, viz., at 7 A.M. and 11 P.M., were close together, show- ing that the stolon had not fallen or risen. Nevertheless, by comparing its position on the morning of the 19th and 21st, it is obvious that the stolon had sunk; and this may be attributed to slow bending down either from its own weight or from geotropism.

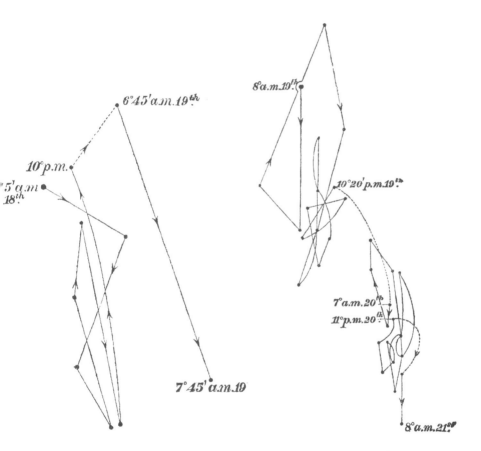

[left] *Fragaria*: circumnutation of stolon, kept in darkness, traced on vertical glass, from 10.45 A.M. May 18th to 7.45 A.M. on 19th.

[right] *Fragaria*: circumnutation of the same stolon as in the last figure, observed in the same manner, and traced from 8 A.M. May 19th to 8 A.M. 21st.

P. Bessa.

Gloriosa superba. Watercolor on vellum by Pancrace Bessa for
Herbier General de l'Amateur.

Gloriosa

FLAME LILY

......

COLCHICACEAE—
COLCHICUM FAMILY

CLIMBING PLANTS

Gloriosa, a genus of climbing plants with eleven species from tropical regions of Africa and Asia, is prized for its dramatically upswept scarlet to orange petals—glorious indeed. Its luscious color and six-stamen one-pistil arrangement of reproductive structures made *Gloriosa* a titillating example for Erasmus Darwin in his poetic (and racy, for the time) showcase of Linnaeus's sexual system of botany, *The Loves of the Plants*, where the flower's stamens and pistil were anthropomorphized into six swains as the "blushing captives" of a wily lady.*[74]

---

* In *The Loves of the Plants*, "Canto I," Erasmus Darwin describes *Gloriosa* as having "Six males, one female. The petals of this beautiful flower with three of the stamens, which are first mature, stand up in apparent disorder; and the pistil bends at nearly a right angle to insert its stigma amongst them. In a few days, as these decline, the other three stamens bend over, and approach the pistil." Poetically, this becomes:

When the young Hours amid her tangled hair
Wove the fresh rose-bud, and the lily fair,
Proud GLORIOSA led three chosen swains,
The blushing captives of her virgin chains.—
—When Time's rude hand a bark of wrinkles spread
Round her weak limbs, and silver'd o'er her head,
Three other youths her riper years engage,
The flatter'd victims of her wily age

The Loves of the Plants is significant in bringing Linnaeus's view of plant sexuality into the cultural mainstream, but also provides a fascinating window into eighteenth century views on gender roles and relationships in society.

Erasmus's grandson Charles would comment on *Gloriosa* pollination too, but in decidedly less exciting terms: "In the *Gloriosa* lily,
the stigma of the grotesque and rectangularly bent pistil is brought,
not into any pathway from the outside towards the nectar secreting
recesses of the flower, but into the circular route which insects follow
in proceeding from one nectary to the other."[75] But the main reason
Darwin grew the plant wasn't so much pollination as its unusual
means of climbing, with tendrils reaching out from the tips of the
leaves. "I am getting very much amused by my tendrils," he opened
a letter to Joseph Hooker in the summer of 1863. He was hoping
Hooker could recommend plants with unusual tendrils, "remarkable
in any way, for development, for odd or peculiar structure or even
for odd place in natural arrangement,"[76] as part of his evolutionary
project, and he soon gratefully received a specimen of fabulous *Gloriosa plantii* (now synonymized with *G. simplex*), which grows widely
in sub-Saharan Africa. He was "amused" and more, excited that this
species provided him with a clue to tendril evolution. Early in his
studies of climbing plants, he struggled with the question of whether
tendrils were derived from stems or leaves and initially leaned toward
the latter as reflected in this note dated 31 January 1864:

> [Bignonia unguis], *with cases of* Gloriosa, [Tropaeolum
> tricolorum] *and* Clematis *makes me strongly suspect that all
> tendrils are first leaf climbers. Good because genesis of tendrils
> otherwise inexplicable.*[77]

He eventually came to see that in some cases (see *Passiflora*, p.
245), flower-peduncles gave rise to tendrils, but in most cases, tendrils were modified leaves. *Gloriosa* and other leaf-tip tendril climbers
(e.g., *Flagellaria indica*, *Uvularia*, and *Nepenthes*) thus represented
an intermediate form in a graduated series, from leaf-climbers
with sensitive petioles to true tendril-bearing plants. "Leaves,"
he concluded in *Climbing Plants*, "may acquire all the leading and

characteristic qualities of tendrils, namely, sensitiveness, spontaneous movement and subsequently increased strength."[78]

> **From:** *The Movements and Habits of Climbing Plants*
> (2nd ed., 1875)

Gloriosa Plantii.—The stem of a half-grown plant continually moved, generally describing an irregular spire, but sometimes oval figures with the longer axes directed in different lines. It either followed the sun, or moved in an opposite course, and sometimes stood still before reversing its direction. One oval was completed in 3 hrs. 40m.; of two horseshoe-shaped figures, one was completed in 4 hrs. 35m. and the other in 3 hrs. The shoots, in their movements, reached points between four and five inches asunder. The young leaves, when first developed, stand up nearly vertically; but by the growth of the axis, and by the spontaneous bending down of the terminal half of the leaf, they soon become much inclined, and ultimately horizontal. The end of the leaf forms a narrow, ribbon-like, thickened projection, which at first is nearly straight, but by the time the leaf gets into an inclined position, the end bends downwards into a well-formed hook. This hook is now strong and rigid enough to catch any object, and, when caught, to anchor the plant and stop the revolving movement. Its inner surface is sensitive, but not in nearly so high a degree as that of the many before-described petioles; for a loop of string, weighing 1.64 grain, produced no effect. When the hook has caught a thin twig or even a rigid fibre, the point may be perceived in from 1 hr. to 3hrs. to have curled a little inwards; and, under favourable circumstances, it curls round and permanently seizes an object in from 8 hrs. to 10 hrs.

The hook when first formed, before the leaf has bent downwards, is but little sensitive. If it catches hold of nothing, it remains open and sensitive for a long time; ultimately the extremity spontaneously and slowly curls inwards, and makes a button-like, flat, spiral coil at the end of the leaf. One leaf was watched, and the hook remained open for thirty-three days; but during the last week the tip had curled so much inwards that only a very thin twig could have been inserted within it. As soon as the tip has curled so much inwards that the hook is converted into a ring, its sensibility is lost; but as long as it remains open some sensibility is retained.

Whilst the plant was only about six inches in height, the leaves, four or five in number, were broader than those subsequently produced; their soft and but little-attenuated tips were not sensitive and did not form hooks; nor did the stem then revolve. At this early period of growth, the plant can support itself; its climbing powers are not required, and consequently are not developed. So again, the leaves on the summit of a full-grown flowering plant, which would not require to climb any higher, were not sensitive and could not clasp a stick. We thus see how perfect is the economy of nature.

Humulus lupulus.   Wild Hops.

E.W. Aug.t 1808.

Humulus lupulus. Watercolor by Elizabeth Wharton, *British Flowers*.

CLIMBING PLANTS

Humulus lupulus, common hops, is a twining perennial dioecious vine that is grown commercially and prized for the aromatic female flower clusters used to preserve and flavor beer.[79] It is also an attractive ornamental that adorns porches and trellises with the aid of myriad tiny hook-like trichomes along the stem that help the vine gain purchase. Of the half dozen *Humulus* species in the genus, native to Europe, southwestern Asia, and North America, *H. lupulus* is certainly the most common, with multiple cultivated varieties. It was practically a staple in the England of Darwin's era; the county of Kent, where Darwin lived, was famous for its hop yards, and he certainly enjoyed a Kentish brew from time to time in his local pubs, the George Inn (now George & Dragon) and Queen's Head, both still serving today. He would surely have sampled, too, some of the fine brews, including "Natural Selection" and "Darwin's Origin," which were produced in 2009 to celebrate the bicentennial of his birth and sesquicentennial of the publication of *On the Origin of Species*.

Hops was the first plant Darwin wrote about in *Movements and Habits of Climbing Plants*—a fine choice, as it's one of the most common representatives of the most common group of climbers he described, "spirally twining plants," which he defined as "those which twine spirally round a support and are not aided by any other movement." *Humulus* is among the small minority of left-handed climbers; viewed from above, they spiral around their support in a clockwise direction, extending from lower-right to upper-left, while the majority of species make right-handed helices, winding counterclockwise.

$$\Bigg[\text{From: } \textbf{\textit{The Movements and Habits of Climbing Plants}} \\ \text{(2nd ed., 1875)} \Bigg]$$

When the shoot of a Hop (*Humulus lupulus*) rises from the ground, the two or three first-formed joints or internodes are straight and remain stationary; but the next-formed, whilst very young, may be seen to bend to one side and to travel slowly round towards all points of the compass, moving, like the hands of a watch, with the sun. The movement very soon acquires its full ordinary velocity. From seven observations made during August on shoots proceeding from a plant which had been cut down, and on another plant during April, the average rate during hot weather and during the day is 2 hrs. 8 m. for each revolution; and none of the revolutions varied much from this rate. The revolving movement continues as long as the plant continues to grow; but each separate internode, as it becomes old, ceases to move.

To ascertain more precisely what amount of movement each internode underwent, I kept a potted plant, during the night and day, in a well-warmed room to which I was confined by illness. A long shoot projected beyond the upper end of the supporting stick and was steadily revolving. I then took a longer stick and tied up the shoot, so that only a very young internode, 1¾ of an inch in length, was left free. This was so nearly upright that its revolution could not be easily observed; but it certainly moved, and the side of the internode which was at one time convex became concave, which, as we shall hereafter see, is a sure sign of the revolving movement. I will assume that it made at least one revolution during the first twenty-four hours. Early the next morning, its position was marked, and it made a second revolution in 9 hrs.; during the latter part of this revolution, it moved much quicker, and the third circle was performed in the evening in a little over 3 hrs. As on the succeeding morning, I found that the shoot revolved in 2 hrs. 45 m.; it must have made during the night four revolutions, each at the average rate of a little over 3 hrs. I should add that the temperature of the room varied only a little. ... From this time forward, the revolutions were easily observed. The thirty-sixth revolution was performed at the usual rate; so was the last or thirty-seventh, but it was not completed; for the internode suddenly became upright, and after moving to the centre, remained motionless. I tied a weight to its upper end, so as to bow it slightly and thus detect any movement; but there was none. Some time before the last revolution was half performed, the lower part of the internode ceased to move.

A few more remarks will complete all that need be said about this internode. It moved during five days; but the more rapid movements, after the performance of the third revolution, lasted during three days and twenty hours. The regular revolutions, from the ninth to thirty-sixth inclusive, were effected at the average rate of 2 hrs. 31 m.; but the weather was cold, and this affected the temperature of the room, especially during the night, and consequently retarded the rate of movement a little. ... After the seventeenth revolution, the internode had grown from 1¾ to 6 inches in length, and carried an internode 1⅞ inch long, which was just perceptibly moving; and this carried a very minute ultimate internode. After the twenty-first revolution, the penultimate internode was 2½ inches long, and probably revolved in a period of about three hours. At the twenty-seventh revolution the lower and still moving internode was 8⅜, the penultimate 3½, and the ultimate 2½ inches in length; and the inclination of the whole shoot was such that a circle 19 inches in diameter was swept by it. When the movement ceased, the lower internode was 9 inches, and the penultimate 6 inches in length; so that, from the twenty-seventh to thirty-seventh revolutions inclusive, three internodes were at the same time revolving.

The lower internode, when it ceased revolving, became upright and rigid; but as the whole shoot was left to grow unsupported, it became after a time bent into a nearly horizontal position, the uppermost and growing internodes still revolving at the extremity, but of course no longer round the old central point of the supporting stick. From the changed position of the centre of gravity of the extremity, as it revolved, a slight and slow swaying movement was given to the long horizontally projecting shoot; and this movement I at first thought was a spontaneous one. As the shoot grew, it hung down more and more, whilst the growing and revolving extremity turned itself up more and more.

With the Hop we have seen that three internodes were at the same time revolving; and this was the case with most of the plants observed by me. With all, if in full health, two internodes revolved; so that by the time the lower one ceased to revolve, the one above was in full action, with a terminal internode just commencing to move.

Bindweed with a purple Flower.

Ipomoea purpurea. Water and bodycolor on vellum by Dame Ann Hamilton,
Drawings of Plants.

PLANT MOVEMENT, CROSS AND SELF-FERTILIZATION

At more than 600 species worldwide, *Ipomoea* is the largest genus in the morning glory family, with a great many beautiful, cultivated varieties adorning the porches, trellises, and gardens of the world. The name is of Greek derivation, inspired by how their twisting, twining habit recalls the winding galleries of bark beetles, a connection reflected in the genus of common bark engraver beetles, *Ips*.

Morning glories were among the many climbing plants Darwin studied. He noted that their growing shoots rotate counterclockwise (against the sun), like most twiners, and later tested their sensitivity to light, defined as *heliotropism*, confirming that, like most twiners, the searching shoots do not respond to a lateral light source. He believed this counterintuitive habit to be adaptive, since gaining height is the most important object for climbers.

[**From: *The Power of Movement in Plants* (1880)**]

Ipomoea caerulea and purpurea.—The leaves on very young plants, a foot or two in height, are depressed at night to between 68° and 80° beneath the horizon; and some hang quite vertically downwards. On the following morning, they again rise into a horizontal position. The petioles become at night downwardly curved, either through their entire length or in the upper part alone; and this apparently causes the depression of the blade.

It seems necessary that the leaves should be well illuminated during the day in order to sleep, for those which stood on the back of a plant before a north-east window did not sleep. ...

As heliotropism is so widely prevalent, and as twining plants are distributed throughout the whole vascular series, the apparent absence of any tendency in their stems to bend towards the light seemed to us so remarkable a fact as to deserve further investigation, for it implies that heliotropism can be readily eliminated. When twining plants are exposed to a lateral light, their stems go on revolving or circumnutating about the same spot, without any evident deflection towards the light; but we thought that we might detect some trace of heliotropism by comparing the average rate at which the stems moved to and from the light during their successive revolutions. Three young plants (about a foot in height) of *Ipomoea caerulea* and four of *I. purpurea*, growing in separate pots, were placed on a bright day before a north-east window in a room otherwise darkened, with the tips of their revolving stems fronting the window. When the tip of each plant pointed directly from the window, and when again towards it, the times were recorded. After a few observations we concluded that we could safely estimate the time taken by each semicircle, within a limit of error of at most 5 minutes. [Twenty-two] semicircles to the light were completed, each on an average in 73.95 minutes; and 22 semicircles from the light each in 73.5 minutes. It may, therefore, be said that they travelled to and from the light at exactly the same average rate; though probably the accuracy of the result was in part accidental. In the evening, the stems were not in the least deflected towards the window.

Movement experiments aside, most of Darwin's work with morning glories was concerned with the beneficial effects of outcrossing. The now-cosmopolitan *Ipomoea purpurea*, a large-flowered species originally from Mexico, was one of the first flowers Darwin worked on to test the effects of cross versus self-fertilization. He grew ten generations over eleven years under netting in his greenhouse to control pollination and growth conditions. His experimental design may not have met modern standards for statistical robustness, but the field was in its infancy then, and his approach did yield useful data.

He found that *I. purpurea* is about as well fertilized with its own pollen (selfed) as with pollen from a different plant, but differences became evident in subsequent generations grown from the resulting seeds. He selected seedlings that germinated at the same time from each group and paired them on opposite sides of pots, giving them sticks to climb in a race to the top. Darwin ultimately found that the cross-pollinated plants tended to be taller and more robust, flower earlier, and produce more seeds than selfed plants. He noted an advantage of cross-pollination in nearly all of his generations, though there was one anomalous selfed plant that bucked the trend and grew a bit taller than even its outcrossed counterparts. Darwin dubbed this especially robust specimen "Hero," and grew out its seeds over three generations to determine if its vigor was heritable. (It was.)

Darwin repeated his multi-generational crossing studies with a host of plants besides the purple morning-glory; foxglove, violets, petunias, corn, and others were dutifully grown out and carefully hand-pollinated generation after generation. Never mathematically inclined, he turned to his cousin Francis Galton to help statistically analyze the heaps of data that resulted and reported the results in *The Effects of Cross and Self Fertilisation*.[80] His hunch from some thirty-seven years earlier, when he first became interested in the effects of crossing, proved correct. "The first and most important of the conclusions which may be drawn from the observations given in this volume," he declared in the summary chapter, "is that generally cross-fertilisation is beneficial, and self-fertilisation often injurious, at least with the plants on which I experimented."[81]

$$\left[\begin{array}{c} \text{From: } \textit{The Effects of Cross and Self Fertilisation} \\ \textit{(2nd ed., 1878)} \end{array} \right]$$

A plant of *Ipomoea purpurea*, or as it is often called in England the convolvulus major, a native of South America, grew in my greenhouse. Ten flowers on this plant were fertilised with pollen from the same flower; and ten other flowers on the same plant were crossed with pollen from a distinct plant. The fertilisation of the flowers with their own pollen was superfluous, as this convolvulus is highly self-fertile; but I acted in this manner to make the experiments correspond in all respects. Whilst the flowers are young, the stigma projects beyond the anthers; and it might have been thought that it could not be fertilised without the aid of humble-bees, which often visit the flowers; but as the flower grows older, the stamens increase in length, and their anthers brush against the stigma, which thus receives some pollen. The number of seeds produced by the crossed and self-fertilised flowers differed very little.

Crossed and self-fertilised seeds obtained in the above manner were allowed to germinate on damp sand, and as often as pairs germinated at the same time they were planted ... on the opposite sides of two pots. Five pairs were thus planted; and all the remaining seeds, whether or not in a state of germination, were planted on the opposite sides of a third pot, so that the young plants on both sides were here greatly crowded and exposed to very severe competition. Rods of iron or wood of equal diameter were given to all the plants to twine up; and as soon as one of each pair reached the summit, both were measured. A single rod was placed on each side of the crowded pot, and only the tallest plant on each side was measured. ...

The mean height of the self-fertilised plants in each of the ten generations is shown in the accompanying diagram, that of the intercrossed plants being taken at 100, and on the right side we see the relative heights of the seventy-three intercrossed plants, and of the seventy-three self-fertilised plants. The difference in height between the crossed and self-fertilised plants will perhaps be best appreciated by an illustration: If all the men

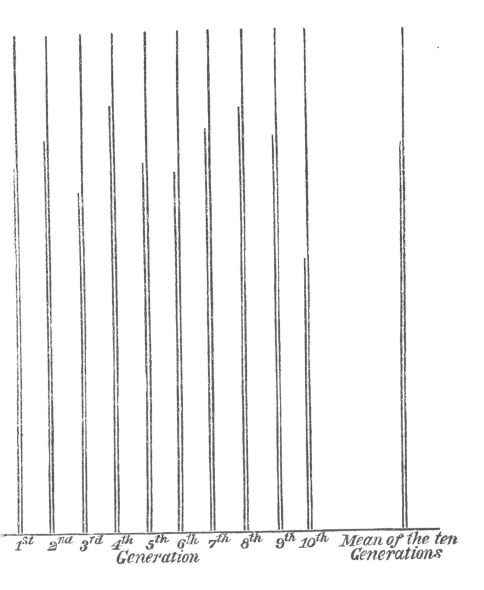

1st   2nd   3rd   4th   5th   6th   7th   8th   9th   10th   Mean of the ten
Generation   Generations

in a country were on an average 6 feet high, and there were some families which had been long and closely interbred, these would be almost dwarfs, their average height during ten generations being only 4 feet 8¼ inches.

It should be especially observed that the average difference between the crossed and self-fertilised plants is not due to a few of the former having grown to an extraordinary height, or to a few of the self-fertilised being extremely short, but to all the crossed plants having surpassed their self-fertilised opponents, with the few following exceptions. The first occurred in the sixth generation, in which the plant named "Hero" appeared; two in the eighth generation, but the self-fertilised plants in this generation were in an anomalous condition, as they grew at first at an unusual rate and conquered for a time the opposed crossed plants; and two exceptions in the ninth generation, though one of these plants only equaled its crossed opponent. Therefore, of the seventy-three crossed plants, sixty-eight grew to a greater height than the self-fertilised plants to which they were opposed.

Painted Lady
Sweet-sented
Peas.

LATHYRUS Siculus flore
odorato Suaverubente.

Lathyrus odoratus. Water and bodycolor on vellum by English School artist,
Album of Garden Flowers.

Lathyrus
SWEET PEA, EVERLASTING
PEA, and RELATIVES

────── ...... ──────

FABACEAE—PEA FAMILY

CLIMBING PLANTS, CROSS AND SELF-FERTILIZATION

Lathyrus is found mainly in temperate regions of the world, cultivated widely for ornament, cover crops, fodder, and food. The best
recognized members of the genus are sweet peas and everlasting
peas, grown since the seventeenth century for their elegant, fragrant
flowers in varied colors. Leaf tendrils are prominent in most species
in the genus, but one, yellow pea (*Lathyrus aphaca*), has stipules that
function as leaves while the leaves have developed into tendrils, and
another, grass vetchling (*L. nissolia*), has grass-like leaves and no
tendrils, prompting Darwin to suggest it may represent reversion to a
primordial ancestor. Darwin grew both native and cultivated species,
and his varied *Lathyrus* studies—rotating movement of seedlings,
cleistogamous flowers, evolution and reversion of tendrils, cross- and
self-pollination, and more—are described in no fewer than five of
his books.

[**From:** *The Movements and Habits of Climbing Plants*
(2nd ed., 1875)]

Nearly all the species of *Lathyrus* possesses tendrils; but *L. nissolia* is
destitute of them. This plant has leaves, which must have struck every
one with surprise who has noticed them, for they are quite unlike those

of all common papilionaceous plants, and resemble those of a grass. In another species, *L. aphaca*, the tendril, which is not highly developed (for it is unbranched, and has no spontaneous revolving-power), replaces the leaves, the latter being replaced in function by large stipules. Now if we suppose the tendrils of *L. aphaca* to become flattened and foliaceous, like the little rudimentary tendrils of the bean, and the large stipules to become at the same time reduced in size, from not being any longer wanted, we should have the exact counterpart of *L. nissolia*, and its curious leaves are at once rendered intelligible to us.

It may be added, as serving to sum up the foregoing views on the origin of tendril-bearing plants, that *L. nissolia* is probably descended from a plant which was primordially a twiner; this then became a leaf-climber, the leaves being afterwards converted by degrees into tendrils, with the stipules greatly increased in size through the law of compensation. After a time, the tendrils lost their branches and became simple; they then lost their revolving-power (in which state they would have resembled the tendrils of the existing *L. nissolia*), and afterwards losing their prehensile power and becoming foliaceous would no longer be thus designated. In this last stage (that of the existing *L. nissolia*) the former tendrils would reassume their original function of leaves, and the stipules which were recently much developed being no longer wanted, would decrease in size. If species become modified in the course of ages, as almost all naturalists now admit, we may conclude that *L. nissolia* has passed through a series of changes, in some degree like those here indicated.

Darwin's most sustained interest in the genus *Lathyrus* was the puzzling prevalence of self-fertilization in cultivated varieties, challenging his belief that species must at least occasionally be cross-fertilized (see *Ipomoea*). Insect visits to their attractive "papilionaceous" flowers, with their colorful paired wing-petal landing pads, rarely led to cross-pollination in Darwin's environs. Darwin's son Francis and his wife Amy reported nectar robbery by bees in *L. japonicus* in Wales.[82] (As it happens, Darwin had collected this very species at Cape Tres Montes in Patagonia[83]—a cross-hemispheric distribution that demonstrates remarkable dispersal abilities.) But nectar robbery could hardly be responsible for the general failure of *Lathyrus* species to cross-pollinate, though when it occurred crossing would be reduced all the more.

In Darwin's experimental plantings, he found that cultivated varieties of the sweet pea remained true even when grown side by side with other varieties. He raised several generations, producing both cross- and self-fertilized plants by hand-pollinating, and confirmed that products of crossing improved in height and vigor— proof that the usual benefits were there when they did cross. When Darwin noticed a leafcutter bee entering a flower in a different way from the bumblebees, triggering the mechanism that deposited pollen on the bee, he suspected that lack of crossing came down to absence of natural pollinators. Writing to Italian botanist Federico Delpino, a frequent correspondent, he wondered if different colored varieties of *L. odoratus* in Italy showed signs of hybridizing if left unprotected from visiting insects. Delpino confirmed that there in its native range, local growers had to plant *L. odoratus* varieties in isolation to prevent intercrossing.[84] Its local pollinator, the leafcutter bee *Megachile ericetorum*, does not occur in Britain.

⌈ **From: *The Effects of Cross and Self Fertilisation in the Vegetable Kingdom*** (2nd ed., 1878) ⌉

Lathyrus odoratus—Almost everyone who has studied the structure of papilionaceous flowers has been convinced that they are specially adapted for cross-fertilisation, although many of the species are likewise capable of self-fertilisation. The case therefore of *Lathyrus odoratus* or the sweet-pea is curious, for in this country it seems invariably to fertilise itself. I conclude that this is so, as five varieties, differing greatly in the colour of their flowers but in no other respect, are commonly sold and come true; yet on inquiry from two great raisers of seed for sale I find that they take no precautions to insure purity—the five varieties being habitually grown close together. I have myself purposely made similar trials with the same result. Although the varieties always come true, yet, as we shall presently see, one of the five well-known varieties occasionally gives birth to another which exhibits all its usual characters. ...

In order to ascertain what would be the effect of crossing two varieties, some flowers on the Purple sweet-pea, which has a dark reddish-purple standard-petal with violet-coloured wing petals and keel, were castrated whilst very young and were fertilised with pollen of the Painted Lady. This latter variety has a pale cherry-coloured standard, with almost white wings and keel. On two occasions, I raised from a flower thus crossed plants perfectly resembling both parent-forms; but the greater number resembled the paternal variety. So perfect was the resemblance that I should have suspected some mistake in the label, had not the plants, which were at first identical in appearance with the father or Painted Lady, later in the season produced flowers blotched and streaked with dark purple. This is an interesting example of partial reversion in the same individual plant as it grows older. The purple-flowered plants were thrown away, as they might possibly have been the product of the accidental self-fertilisation of the mother-plant, owing to the castration not having been effectual. But the plants which resembled in the colour of their flowers the paternal variety or Painted Lady were preserved, and their seeds saved. Next summer many plants were raised from these seeds, and they generally resembled their grandfather the Painted Lady, but most of them had their wing-petals streaked and stained with dark pink; and a few had pale purple wings with the standard of a darker crimson than is natural to the Painted Lady, so that they formed a new sub-variety. Amongst these plants, a single one appeared having purple flowers like those of the grandmother, but with the petals slightly streaked with a paler tint: this was thrown away. Seeds were again saved from the foregoing plants and the seedlings thus raised still resembled the Painted Lady, or great-grandfather; but they now varied much, the standard petal varying from pale to dark red, in a few instances with blotches of white; and the wing-petals varied from nearly white to purple, the keel being in all nearly white.

As no variability of this kind can be detected in plants raised from seeds, the parents of which have grown during many successive generations in close proximity, we may infer that they cannot have intercrossed. What does occasionally occur is that in a row of plants raised from seeds of one variety, another variety true of its kind appears; for instance, in a long row of scarlets (the seeds of which had been carefully gathered from Scarlets for the sake of this experiment) two Purples and one Painted Lady appeared. Seeds from these three aberrant plants were saved and sown in separate beds. The seedlings from both the Purples were chiefly Purples, but with some Painted Ladies and some Scarlets. The seedlings from the aberrant Painted Lady were chiefly Painted Ladies with some Scarlets. Each variety, whatever its percentage may have been, retained all its characters perfect, and there was no streaking or blotching of the colours, as in the foregoing plants of crossed origin. ...

From the evidence now given, we may conclude that the varieties of the sweet-pea rarely or never intercross in this country; and this is a highly remarkable fact, considering, firstly, the general structure of the flowers; secondly, the large quantity of pollen produced, far more than is requisite for self-fertilisation; and thirdly, the occasional visits of insects. That insects should sometimes fail to cross-fertilise the flowers is intelligible, for I have thrice seen humble-bees of two kinds, as well as hive-bees, sucking the nectar, and they did not depress the keel-petals so as to expose the anthers and stigma; they were therefore quite inefficient for fertilising the flowers. One of these bees, namely, *Bombus lapidarius*, stood on one side at the base of the standard and inserted its proboscis beneath the single separate stamen, as I afterwards ascertained by opening the flower and finding this stamen prised up. Bees are forced to act in this manner from the slit in the staminal tube being closely covered by the broad membranous margin of the single stamen, and from the tube not being perforated by nectar-passages. On the other hand, in the three British species of *Lathyrus* which I have examined, and in the allied genus *Vicia*, two nectar-passages are present. Therefore, British bees might well be puzzled how to act in the case of the sweet-pea. ... One of my sons caught an elephant sphinx-moth whilst visiting the flowers of the sweet-pea, but this insect would not depress the wing-petals and keel. On the other hand, I have seen on one occasion hive-bees, and two or three occasions the *Megachile willughbiella* in the act of depressing the keel; and these bees had the under sides of their bodies thickly covered with pollen and could not thus fail to carry pollen from one flower to the stigma of another. Why then do not the varieties occasionally intercross, though this would not often happen, as insects so rarely act in an efficient manner? ... Whatever the cause may be, we may conclude, that in England the varieties never or very rarely intercross. But it does not follow from this that they would not be crossed by the aid of other and larger insects in their native country, which in botanical works is said to be the south of Europe and the East Indies. Accordingly, I wrote to Professor Delpino, in Florence, and he informs me "that it is the fixed opinion of gardeners there that the varieties do intercross, and that they cannot be preserved pure unless they are sown separately."

Antirrhinum Linaria.

Linaria vulgaris. Hand-colored engraved plate from *Flora Londinensis*, published by William Curtis.

Linaria

TOADFLAX

———— ...... ————

PLANTAGINACEAE—
PLANTAIN FAMILY

CROSS AND SELF-FERTILIZATION

Some 150 species are classified in the genus *Linaria*, a temperate
European, African, and Asian group in a family related to snapdrag-
ons. The Latin name, and the latter half of the common name, refers
to the similarity of the leaves of those of many flax plants (see p. 121).
Darwin grew several species of *Linaria* in his experimental gardens,
mainly as part of his research into the effects of self-fertilization
versus outcrossing in flowering plants. One set of studies involved
common toadflax, *Linaria vulgaris*, a European species now with a
cosmopolitan distribution, also called butter-and-eggs due to its
lovely yellow and orange blossoms. In the course of his investigations into the role of insects as pollen-carrying go-betweens, Darwin
found that without insect aid, the flowers of *L. vulgaris* are sterile for
the most part, with minimal seed production; 100 seeds yielded by
cross-fertilized flowers compared to just 14 seeds from self-pollinated
ones. What's more, the seedlings grown from the few seeds that were
produced by self-pollination tended to be shorter and less vigorous
than those resulting from outcrossed flowers.

From: *The Effects of Cross and Self Fertilisation in the Vegetable Kingdom* (2nd ed., 1878)

It often occurred to me that it would be advisable to try whether seedlings from cross-fertilised flowers were in any way superior to those from self-fertilised flowers. But as no instance was known with animals of any evil appearing in a single generation from the closest possible interbreeding, that is between brothers and sisters, I thought that the same rule would hold good with plants; and that it would be necessary at the sacrifice of too much time to self-fertilise and intercross plants during several successive generations, in order to arrive at any result. I ought to have reflected that such elaborate provisions favouring cross-fertilisation, as we see in innumerable plants, would not have been acquired for the sake of gaining a distant and slight advantage, or of avoiding a distant and slight evil. Moreover, the fertilisation of a flower by its own pollen corresponds to a closer form of interbreeding than is possible with ordinary bi-sexual animals; so that an earlier result might have been expected.

I was at last led to make the experiments recorded in the present volume from the following circumstance. For the sake of determining certain points with respect to inheritance, and without any thought of the effects of close interbreeding, I raised close together two large beds of self-fertilised and crossed seedlings from the same plant of *Linaria vulgaris*. To my surprise, the crossed plants when fully grown were plainly taller and more vigorous than the self-fertilised ones. Bees incessantly visit the flowers of this *Linaria* and carry pollen from one to the other; and if insects are excluded, the flowers produce extremely few seeds; so that the wild plants from which my seedlings were raised must have been intercrossed during all previous generations. It seemed therefore quite incredible that the difference between the two beds of seedlings could have been due to a single act of self-fertilisation; and I attributed the result to the self-fertilised seeds not having been well ripened, improbable as it was that all should have been in this state, or to some other accidental and inexplicable cause. ...

The trial was afterwards repeated with more care; but as this was one of the first plants experimented on, my usual method was not followed. Seeds were taken from wild plants growing in this neighbourhood and sown in poor soil in my garden. Five plants were covered with a net, the others being left exposed to the bees, which incessantly visit the flowers of this species, and which, according to H. Müller, are the exclusive fertilisers, This excellent observer remarks that, as the stigma lies between the anthers and is mature at the same time with them, self-fertilisation is possible. But so few seeds are produced by protected plants, that the pollen and stigma of the same flower seem to have little power of mutual interaction. The exposed plants bore numerous capsules forming solid spikes. Five of these capsules were examined and appeared to contain an equal number of seeds; and these being counted in one capsule were found to be 166. The five protected plants produced altogether only twenty-five capsules, of which five were much finer than all the others, and these contained an average of 23.6 seeds, with a maximum in one capsule of fifty-five. So that the number of seeds in the capsules on the exposed plants to the average number in the finest capsules on the protected plants was as 100 to 14.

Some of the spontaneously self-fertilised seeds from under the net, and some seeds from the uncovered plants naturally fertilised and almost certainly intercrossed by the bees, were sown separately in two large pots of the same size; so that the two lots of seedlings were not subjected to any mutual competition. Three of the crossed plants when in full flower were measured, but no care was taken to select the tallest plants; their heights were $7^{4}/_{8}$, $7^{2}/_{8}$, and $6^{4}/_{8}$ inches; averaging 7.08 in height. The three tallest of all the self-fertilised plants were then carefully selected, and their heights were $6^{3}/_{8}$, $5^{5}/_{8}$, and $5^{2}/_{8}$, averaging 5.75 in height. So that the naturally crossed plants were to the spontaneously self-fertilised plants in height, at least as much as 100 to 81.

Modern studies show that the largely self-incompatible flowers of *L. vulgaris* employ several strategies to command the attention of pollinators, including nectaries (that conspicuous spur where nectar collects) and olfactory and visual attractants.[85]

Darwin also observed bees visiting the beautiful purple toadflax (*L. purpurea*), a tall species native to the Italian peninsula and now grown in gardens worldwide. The uniform coloring of these flowers cast doubt, for him, on the suggestion made by German botanist Konrad Sprengel in 1793 that streaks and stripes on petals played a role in attracting or directing insect visitors, facilitating pollination. Darwin remarked in a letter published in the *Gardeners' Chronicle* that he knew "hardly any flower which bees open and insert their proboscis into, more rapidly, than the common tall linaria, which has a little purplish well-closed flower; I have watched one humble-bee suck twenty-four flowers in one minute, yet on this flower there are no streaks of colour to guide these quick and clever workmen."[86]

Sprengel's hypothesis is largely accepted today, with such streaks and stripes now termed "nectar guides," and an understanding that many flowers which appear uniform in color to our eyes actually do sport them, but in the ultraviolet end of the spectrum that bees can perceive. Darwin would be fascinated to learn that nectar guides may in fact reduce the nectar-robbery he often observed bumble bees engaging in, by reducing the bees' handling time and access to the business end of a flower—a double benefit to the plant, since reduced nectar robbery translates into increased pollination.[87]

Linum perenne. Water and bodycolor on vellum by Dame Ann Hamilton,
Drawings of Plants.

Linum

FLAX

......

LINACEAE—FLAX FAMILY

Among the more than 200 species of *Linum*, the largest genus in the flax or linseed family, many are cultivated for their beautiful flowers while others are grown for food and fiber. The common flax *Linum usitatissimum*, a species cultivated since antiquity, is so versatile that its very name, *usitatissimum,* is thought to derive from the Latin meaning "most useful." *Linum* was most useful to Darwin as a nice illustration of floral dimorphism, found in about half of *Linum* species, where one plant type, or morph, bears long-styled pistils and short stamens, and another morph has short-styled pistils and long stamens. Darwin was the first naturalist to recognize the functional (and adaptive) significance of this phenomenon, termed *heterostyly*. He had first discovered flower dimorphism in primroses (see p. 171), but found *Linum* to be an even better example.

When he became aware of *Linum* floral dimorphism in 1861, Darwin initiated a series of crossing experiments with the species *L. grandiflorum* and *L. perenne*, procured with the help of his friends at the Royal Botanic Gardens, Kew. The results, first presented in a paper to the Linnean Society of London[88] and later incorporated into *Forms of Flowers*, proved as puzzling as they were interesting. As far as he could make out with a microscope, pollen from the two morphs were identical. Yet, the short-style morph proved fertile with pollen from either morph, while flowers of the long-style morph could only be fertilized with pollen from short-style plants. He called fertile matches "legitimate" crosses, and infertile ones "illegitimate." Noting that "it may be said that the two pollens and the two stigmas mutually recognise each other by some means,"[89] Darwin correctly inferred that this recognition system functioned to prevent self-fertilization and promote outcrossing, presaging the discovery in the next century of genetic and biochemical self-recognition mechanisms in plants—in particular the widespread self-incompatibility (SI) genetic system.* His results prompted an analysis of the differences between wind and insect pollination.

[**From: *The Different Forms of Flowers on Plants of the Same Species* (1877)**]

It has long been known that several species of *Linum* present two forms, and, having observed this fact in *L. flavum* more than thirty years ago, I was led, after ascertaining the nature of heterostylism in *Primula*, to examine the first species of *Linum* which I met with, namely, the beautiful *L. grandiflorum*. This plant exists under two forms, occurring in about equal numbers, which differ little in structure but greatly in function. The foliage, corolla, stamens, and pollen-grains (the latter examined both distended with water and dry) are alike in the two forms. ... The difference is confined to the pistil; in the short-styled form the styles and the stigmas are only about half the length of those in the long-styled. A more important distinction is that the five stigmas in the short-styled form diverge greatly from one another and pass out between the filaments of the stamens, and thus lie within the tube of the corolla. In the long-styled form, the elongated stigmas stand nearly upright and alternate with the anthers. In this latter form, the length of the stigmas varies considerably, their upper extremities projecting even

---

* *Self-incompatibility* (SI) is a general term for several genetic mechanisms that prevent self-fertilization (and thus promote outcrossing) in flowering plants and a few other groups.

a little above the anthers or reaching up only to about their middle. Nevertheless, there is never the slightest difficulty in distinguishing between the two forms; for, besides the difference in the divergence of the stigmas, those of the short-styled form never reach even to the bases of the anthers. In this form, the papillae on the stigmatic surfaces are shorter, darker-coloured, and more crowded together than in the long-styled form; but these differences seem due merely to the shortening of the stigma, for in the varieties of the long-styled form with shorter stigmas, the papillae are more crowded and darker-coloured than in those with the longer stigmas. Considering the slight and variable differences between the two forms of this *Linum*, it is not surprising that hitherto they have been overlooked. ...

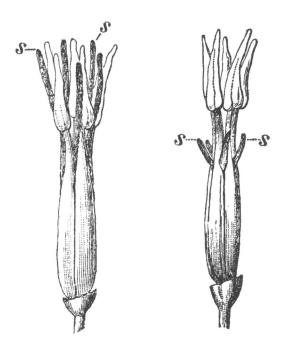

Linum grandiflorum. Long-styled form, left. Short-styled form, right. *s* = stigma

The absolute sterility (judging from the experiments of 1861) of the long-styled plants with their own-form pollen led me to examine into its apparent cause; and the results are so curious that they are worth giving in detail. The experiments were tried on plants grown in pots and brought successively into the house.

Pollen from a short-styled plant was placed on the five stigmas of a long-styled flower, and these, after thirty hours, were found deeply pene-trated by a multitude of pollen-tubes, far too numerous to be counted; the stigmas had also become discoloured and twisted. I repeated this experiment on another flower, and in eighteen hours the stigmas were penetrated by a multitude of long pollen-tubes. This is what might have been expected, as the union is a legitimate one. The converse experiment was likewise tried, and pollen from a long-styled flower was placed on the stigmas of a short-styled flower, and in twenty-four hours the stigmas were discoloured, twisted, and penetrated by numerous pollen-tubes; and this, again, is what might have been expected, as the union was a legitimate one. ...

I could add other experiments; but those now given amply suffice to show that the pollen-grains of a short-styled flower placed on the stigma of a long-styled flower emit a multitude of tubes after an interval of from five to six hours and penetrate the tissue ultimately to a great depth; and that after twenty-four hours the stigmas thus penetrated change colour, become twisted, and appear half-withered. On the other hand, pollen-grains from a long-styled flower placed on its own stigmas, do not emit their tubes after an interval of a day, or even three days; or at most only three or four grains out of a multitude emit their tubes, and these apparently never penetrate the stigmatic tissue deeply, and the stigmas themselves do not soon become discoloured and twisted.

The plants both of *L. perenne and grandiflorum* grew with their branches interlocked, and with scores of flowers of the two forms close together; they were covered by a rather coarse net, through which the wind, when high, passed; and such minute insects as Thrips could not, of course, be excluded; yet we have seen that the utmost possible amount of accidental fertilisation on seventeen long-styled plants in the one case, and on eleven long-styled plants in the other, resulted in the production, in each case, of three poor capsules; so that when the proper insects are excluded, the wind does hardly anything in the way of carrying pollen from plant to plant. I allude to this fact because botanists, in speaking of the fertilisation of various flowers, often refer to the wind or to insects as if the alternative were indifferent. This view, according to my experience, is entirely erroneous. When the wind is the agent in carrying pollen, either from one sex to the other, or from hermaphrodite to hermaphrodite, we can recognise structure as manifestly adapted to its action as to that of insects when these are the carriers. We see adaptation to the wind in the incoherence of the pollen,—in the inordinate quantity produced (as in the Coniferae, Spinage, &c.),—in the dangling anthers well fitted to shake out the pollen,—in the absence or small size of the perianth,—in the protrusion of the stigmas at the period of fertilisation,—in the flowers being produced before they are hidden by the leaves,—and in the stigmas being downy or plumose (as in the Gramineae, Docks, &c.), so as to secure the chance-blown grains. In plants which are fertilised by the wind, the flowers do not secrete nectar, their pollen is too incoherent to be easily collected by insects, they have not bright-coloured corollas to serve as guides, and they are not, as far as I have seen, visited by insects. When insects are the agents of fertilisation (and this is incomparably the more frequent case with hermaphrodite plants), the wind plays no part, but we see an endless number of adaptations to ensure the safe transport of the pollen by the living workers.

The great blue Lupine.

Lupinus pilosus. Water and bodycolor on vellum by English School artist,
Album of Garden Flowers.

PLANT MOVEMENT

Lupines are one of a veritable pack of plants whose names make reference to wolves—from the Latin word for wolf, *lupus*. In this case, the name may reference the palmately divided leaves, evoking a wolf's paw, or the ancient (and misguided) belief that these plants are ravenous as wolves in depleting soils of nutrients. Many are found in disturbed habitat with infertile soils, but their effect is quite the opposite of depletion—these nitrogen-fixing legumes help restore barren soils.

The genus is expansive—more than 200 species are found from Europe to north Africa and Eurasia and throughout the Americas (the center of lupine diversity). While there are a few woody lupines, most are herbaceous annuals and perennials of open fields and meadows. Many species are toxic, but others have been used in many ways by people; some have been grown since antiquity for their edible seeds, others are planted as livestock fodder or cover crops. A multitude of ornamental varieties have been produced, prized for their large spikes of brightly colored flowers. In Darwin's time, lupines were a mainstay of the English garden, and it was in the gardens of his childhood home in Shrewsbury and nearby Maer, home of the Wedgwoods, his wife Emma's family, that he began observing pollination of lupines and other flowers in the summers of 1840 and 1841, flush with the excitement of his recent conversion to the heretical idea of species change (known as transmutation in his day and evolution today).[90]

Darwin came to learn that lupine stamens are distinctive, with five large colorful sagittate anthers and five smaller and differently colored ones, a curious dimorphism that quickly caught his imagination. He had first read about this in an 1841 treatise by Swiss pastor and botanist Jean Pierre Étienne Vaucher and wondered if it was another adaptation to promote outcrossing, like heterostyly. How widespread was this phenomenon? "Can you think of plants which have differently coloured anthers or pollen in same flowers," he asked Joseph Hooker in August 1862. "It would be a safe guide to dimorphism.—Do just think of this."[91] Hooker was able to provide him with a list of cases. Darwin threw himself into crossing experiments, including among his many experimental plants two lupine species: common yellow and blue lupines, *Lupinus luteus* and *L. pilosus*. As described in *Cross and Self Fertilisation*, his lupines developed fruits and seeds freely whether crossed or self-pollinated, but cross-pollination trials led to a much more vigorous second generation.

Darwin returned to lupines in his later studies of movement in plants. He found that the leaves of some species trace extraordinary ellipses (circumnutation) in the course of the day but do not "sleep" (nyctitropism) at night. Others do sleep in various ways, moving up or down, or rotating from a horizontal to a vertical attitude, their many-pointed leaves aptly likened by him to so many vertically hung "stars" at night. Ever attuned to evolutionary transitions, Darwin recognized that yellow lupine was a one-plant showcase for diversified nyctitropic behavior. He wrote, "Four leaves on the same plant, which had their leaflets horizontal at noon, formed vertical stars at night; and three other leaves equally horizontal at noon, had all their leaflets sloping downwards at night. So that the leaves on this one plant assumed at night three different positions. Though we cannot account for this fact, we can see that such a stock might readily give birth to species having widely different nyctitropic habits."[92]

[From: *The Power of Movement in Plants* (1880)]

Lupinus —The palmate or digitate leaves of the species in this large genus sleep in three different manners. One of the simplest is that all the leaflets become steeply inclined downwards at night, having been during the day extended horizontally. This is shown in the accompanying figures, of a leaf of *L. pilosus*, as seen during the day from vertically above, and of another leaf asleep with the leaflets inclined downwards. As in this position they are crowded together, and as they do not become folded like those in the genus Oxalis, they cannot occupy a vertically dependent position; but they are often inclined at an angle of 50° beneath the horizon. In this species, whilst the leaflets are sinking, the petioles rise up, in two instances when the angles were measured to the extent of 23°. The leaflets of *L. sub-carnosus* and *arboreus*, which were horizontal during the day, sank down at night in nearly the same manner; the former to an angle of 38° and the latter of 36° beneath the horizon; but their petioles did not move in any plainly perceptible degree. It is, however, quite possible, as we shall

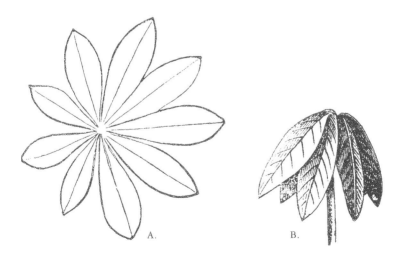

A. B.

Lupinus pilosus: A, leaf seen from vertically above in daytime;
B, leaf asleep, seen laterally at night.

presently see, that if a large number of plants of the three foregoing and of the following species were to be observed at all seasons, some of the leaves would be found to sleep in a different manner.

In the two following species, the leaflets, instead of moving downwards, rise at night. With *L. hartwegii*, some stood at noon at a mean angle of 36° above the horizon, and at night at 51°, thus forming together a hollow cone with moderately steep sides. The petiole of one leaf rose 14° and of a second 11° at night. With *L. luteus*, a leaflet rose from 47° at noon to 65° above the horizon at night, and another on a distinct leaf rose from 45° to 69°. The petioles, however, sink at night to a small extent, viz., in three instances by 2°, 6°, and 9° 30 seconds. Owing to this movement of the petioles, the outer and longer leaflets have to bend up a little more than the shorter and inner ones, in order that all should stand symmetrically at night. ...

We now come to a remarkable position of the leaves when asleep, which is common to several species of Lupines. On the same leaf, the shorter leaflets, which generally face the centre of the plant, sink at night, whilst the longer ones on the opposite side rise; the intermediate and lateral ones merely twisting on their own axes. But there is some variability with respect to which leaflets rise or fall. As might have been expected from such diverse and complicated movements, the base of each leaflet is developed (at least in the case of *L. luteus*) into a pulvinus. The result is that all the leaflets on the same leaf stand at night more or less highly inclined, or even quite vertically, forming in this latter case a vertical star. This occurs with the leaves of a species purchased under the name of *L. pubescens*; and in the accompanying figures we see at A the leaves in their diurnal position; and at B the same plant at night with the two upper leaves having their leaflets almost vertical. At C another leaf, viewed laterally, is shown with the leaflets quite vertical. It is chiefly or exclusively the youngest leaves which form at night vertical stars. But there is much variability in the position of the leaves at night on the same plant; some remaining with their leaflets almost horizontal, others forming more or less highly inclined or vertical stars, and some with *all* their leaflets sloping downwards, as in our first class of cases. It is also a remarkable fact, that although all the plants produced from the same lot of seeds were identical in appearance, yet some individuals at night had the leaflets of all their leaves arranged so as to form more or less highly inclined stars; others had them all sloping downwards and never forming a star; and others, again, retained them either in a horizontal position or raised them a little.

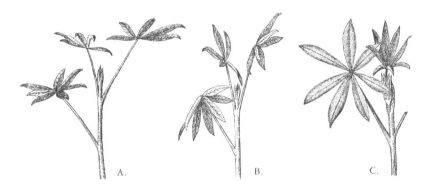

Lupinus pubescens: A, leaf viewed laterally during the day; B, same leaf at night; C, another leaf with the leaflet forming a vertical star at night.

Maurandya scandens. Hand-colored engraving, drawn by Sydenham Edwards, from *The Botanical Magazine*, 13: 460.

CREEPING or TRAILING SNAPDRAGON

——— ...... ———

PLANTAGINACEAE—PLANTAIN FAMILY

CLIMBING PLANTS

Maurandya is a genus of four perennial vine species found from the southwestern United States to Central America, known for their "scandent" (clambering or sprawling) climbing habit. In his work on climbing plants, Darwin discussed two species, *Maurandya barclayana* and *M. scandens* (*semperflorens* in Darwin's era), characterized by clambering that is facilitated by twining leaf petioles and flower peduncles. These plants are closely related to two other related genera of Central American climbers that Darwin wrote about—*Rhodochiton*, consisting of three species with pendant flowers on elongate peduncles and an inflated calyx, and *Lophospermum*, a genus of seven species with flowers that are horizontal to ascending and a calyx that is not inflated.

   In the second edition of his book on climbing plants, Darwin noted that the young flower-peduncles of *M. scandens* revolve in a tight circular motion and respond slightly to gentle rubbing by slowly bending in the direction of the touch stimulus. He concluded that the plant "certainly does not profit by these two feebly developed powers," but saw adaptive potential, speculating that they might easily evolve into full-fledged tendrils. "We see at least that the *Maurandia* [sic] might, by a little augmentation of the powers which it already possesses, come first to grasp a support by its flower-peduncles, and then, by the abortion of some of its flowers (as with *Vitis* or *Cardiospermum*), acquire perfect tendrils."[93]

[**From:** *The Movements and Habits of Climbing Plants*
(2nd ed., 1875)

Maurandia Barclayana.—A thin, slightly bowed shoot made two revolutions, following the sun, each in 3 hrs. 17 min.; on the previous day this same shoot revolved in an opposite direction. The shoots do not twine spirally but climb excellently by the aid of their young and sensitive petioles. These petioles, when lightly rubbed, move after a considerable interval of time, and subsequently become straight again. A loop of thread weighing ⅛th of a grain caused them to bend.

Maurandia semperflorens.—This freely growing species climbs exactly like the last, by the aid of its sensitive petioles. A young internode made two circles, each in 1 hr. 46 min.; so that it moved almost twice as rapidly as the last species. The internodes are not in the least sensitive to a touch or pressure. I mention this because they are sensitive in a closely allied genus, namely, *Lophospermum.* The present species is unique in one respect. Mohl asserts that "the flower-peduncles, as well as the petioles, wind like tendrils;" but he classes as tendrils such objects as the spiral flower-stalks of the *Vallisneria.* This remark, and the fact of the flower-peduncles being decidedly flexuous, led me carefully to examine them. They never act as true tendrils; I repeatedly placed thin sticks in contact with young and old peduncles, and I allowed nine vigorous plants to grow through an entangled mass of branches; but in no one instance did they bend round any object. It is indeed in the highest degree improbable that this should occur, for they are generally developed on branches which have already securely clasped a support by the petioles of their leaves; and when borne on a free depending branch, they are not produced by the terminal portion of the internode which alone has the power of revolving; so that they could be brought only by accident into contact with any neighbouring object. Nevertheless (and this is the remarkable fact) the flower-peduncles, whilst young, exhibit feeble revolving powers, and are slightly sensitive to a touch. Having selected some stems which had firmly clasped a stick by their petioles, and having placed a bell-glass over them, I traced the movements of the young flower-peduncles. The tracing generally formed a short and extremely irregular line, with little loops in its course. A young peduncle 1½ inch in length was carefully observed during a whole day, and it made four and a half narrow, vertical, irregular, and short ellipses—each at an average rate of about 2 hrs. 25 m. An adjoining peduncle described during the same time similar, though fewer, ellipses. As the plant had occupied for some time exactly the same position, these movements could not be attributed to any change in the action of the light. Peduncles, old enough for the coloured petals to be just visible, do not move. With respect to irritability, I rubbed two young peduncles (1½ inch in length) a few times very lightly with a thin twig; one was rubbed on the upper, and the other on the lower side, and they became in between 4 hrs. and 5 hrs. distinctly bowed towards these sides; in 24 hrs. subsequently, they straightened themselves. Next day they were rubbed on the opposite sides, and they became perceptibly curved towards these sides. Two other and younger peduncles (three-fourths of an inch in length) were lightly rubbed on their adjoining sides, and they became so much curved towards one another that the arcs of the bows stood at nearly right angles to their previous direction; and this was the greatest movement seen by me. Subsequently they straightened themselves. Other peduncles, so young as to be only three-tenths of an inch in length, became curved when rubbed.

In the nine vigorous plants observed by me, it is certain that neither the slight spontaneous movements nor the slight sensitiveness of the flower-peduncles aided the plants in climbing. If any member of the Scrophulariaceae had possessed tendrils produced by the modification of flower-peduncles, I should have thought that this species of *Maurandia* had perhaps retained a useless or rudimentary vestige of a former habit; but this view cannot be maintained. We may suspect that, owing to the principle of correlation, the power of movement has been transferred to the flower-peduncles from the young internodes, and sensitiveness from the young petioles. But to whatever cause these capacities are due, the case is interesting; for, by a little increase in power through natural selection, they might easily have been rendered as useful to the plant in climbing, as are the flower-peduncles of *Vitis or Cardiospermum*.

About this last observation, Asa Gray wrote to Darwin in December 1875, "I see a little matter in which I can help you out. On p. 198, you say of *Maurandia* that, with little more ado, it might grasp a support by its flower-peduncles. Well, I think it does that sometimes." He thought that was true of *M. antirrhiniflora*, for one, and was sure that the two California species *M. stricta* and *M. cooperi* had touch-sensitive peduncles as well. Darwin thanked him, responding, "What would I not have given for them when I was preparing the new Edit; but it is now too late, for I do not suppose I shall ever again touch the book."[94] He did in fact bring out a final edition, in 1882, but edits were minimal, and he left the discussion of *Maurandya* unchanged.

Humble Plant.

Mimosa pudica. Water and bodycolor on vellum by Dame Ann Hamilton,
Drawings of Plants.

```
┌─────────────────────────────────────────┐
│                                           │
│              *Mimosa*                      │
│          SENSITIVE PLANT                  │
│          ─────  ......  ─────             │
│         FABACEAE—PEA FAMILY               │
│                                           │
└─────────────────────────────────────────┘
```

PLANT MOVEMENT

Mimosa was named by Carl Linnaeus from the Spanish *mimoso*, meaning "sensitive." It is a large genus native to the tropical regions of the world, most abundant in the American tropics. *Mimosa sensitiva*, from Brazil, was introduced to Europe in the seventeenth century as an object of wonder for its touch-sensitive leaves. Darwin's grandfather Erasmus Darwin, the noted physician and poet, romanticized (and anthropomorphized) the sensitive plant in verse as a shy damsel in *The Loves of the Plants*: "Weak with nice sense, this chaste Mimosa stands / From each rude touch withdraws her timid hands;"[95]

Some forty years later, his grandson Charles saw it while on the *Beagle* voyage, as mentioned in a letter to Joseph Hooker: "By the way I well remember in Brazil walking through a great bed of *Mimosa sensitiva* and leaving a track as if an elephant had passed that way. The plant has interested me ever since."[96]

M. pudica, easier to grow, became much more popular as a curiosity for its immediate shrinking response to shock or touch. Both species and many others have turgid pulvini that enable the leaves to retract and fold on themselves not only by touch, but at nightfall. "The whole appearance of many plants is wonderfully changed at night, as may be seen with Oxalis and still more plainly with Mimosa," Darwin wrote in the concluding section of *Movement*.[97] With the help of his son William, he tested *M. pudica* and *M. albida* for reactions to water and chemicals, though he did not publish those results. It was the circular (circumnutation) and nocturnal (nyctitropic) movement that most fascinated him, leading him to experiment on both cotyledons and mature plants, tracing the slow movement of stems and leaves, including the petioles, pinnae, and leaflets, on glass plates over many hours to illustrate their long, looping movements.

[**From: *The Power of Movement in Plants* (1880)**]

Mimosa pudica—This plant has been the subject of innumerable observations; but there are some points in relation to our subject which have not been sufficiently attended to. At night, as is well known, the opposite leaflets come into contact and point towards the apex of the leaf; they thus become neatly imbricated with their upper surfaces protected. The four pinnae also approach each other closely, and the whole leaf is thus rendered very compact. The main petiole sinks downwards during the day till late in the evening and rises until very early in the morning. The stem is continually circumnutating at a rapid rate, though not to a wide extent. Some very young plants, kept in darkness, were observed during two days, and although subjected to a rather low temperature of 57°–59° F., the stem of one described four small ellipses in the course of 12 h. We shall immediately see that the main petiole is likewise continually circumnutating, as is each separate pinna and each separate leaflet. Therefore, if the movement of the apex of any one leaflet were to be traced, the course described would be compounded of the movements of four separate parts.

A filament had been fixed on the previous evening, longitudinally to the main petiole of a nearly full-grown, highly-sensitive leaf (four inches in length), the stem having been secured to a stick at its base; and a tracing was made on a vertical glass in the hot-house under a high temperature. In the figure given, the first dot was made at 8.30 A.M. August 2nd, and the last at 7 P.M. on the 3rd. During 12 h. on the first day, the petiole moved thrice downwards and twice upwards. Within the same length of time on the second day, it moved five times downwards and four times upwards. As the ascending and descending lines do not coincide, the petiole manifestly circumnutates; the great evening fall and nocturnal rise being an exaggeration of one of the circumnutations. It should, however, be observed that the petiole fell much lower down in the evenings than could be seen on the vertical glass or is represented in the diagram. After 7 P.M. on the 3rd (when the last dot in [the figure] was made) the pot was carried into a bed-room,

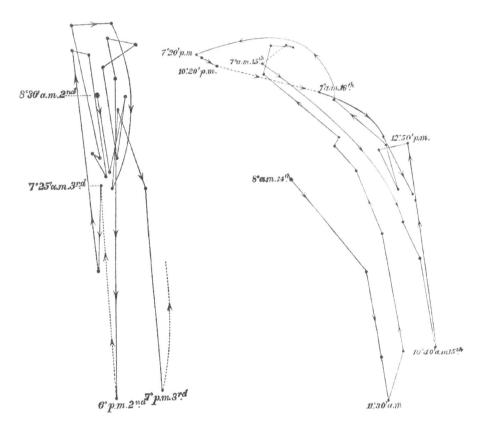

Mimosa pudica: circumnutation and nyctitropic movement of main petiole [left] and circumnutation and nyctitropic movement of a leaflet (with pinna secured) [right], traced on a vertical glass.

and the petiole was found at 12.50 A.M. (i.e. after midnight) standing almost upright, and much more highly inclined than it was at 10.40 P.M. When observed again at 4 A.M. it had begun to fall, and continued falling till 6.15 A.M., after which hour it zigzagged and again circumnutated. Similar observations were made on another petiole, with nearly the same result. ...

It has also been stated that each separate leaflet circumnutates. A pinna was cemented with shellac on the summit of a little stick driven firmly into the ground, immediately beneath a pair of leaflets, to the midribs of both of which excessively fine glass filaments were attached. This treatment did not injure the leaflets, for they went to sleep in the usual manner, and long retained their sensitiveness. the movements of one of them were traced during 49 h., as shown. On the first day, the leaflet sank down till 11.30 A.M., and then rose till late in the evening in a zigzag line, indicating circumnutation. On the second day, when more accustomed to its new state, it oscillated twice up and twice down during the 24 h. This plant was subjected to a rather low temperature, viz., 62°–64° F.; had it been kept warmer, no doubt the movements of the leaflet would have been much more rapid and complicated. It may be seen in the diagram that the ascending and descending lines do not coincide; but the large amount of lateral movement in the evening is the result of the leaflets bending towards the apex of the leaf when going to sleep. ...

Mimosa albida.—The leaves of this plant, one of which is here figured ... present some interesting peculiarities. It consists of a long petiole bearing only two pinnae (here represented as rather more divergent than is usual), each with two pairs of leaflets. But the inner basal leaflets are greatly reduced in size, owing probably to the want of space for their full development, so that they may be considered as almost rudimentary. They vary somewhat in size, and both occasionally disappear, or only one. Nevertheless, they are not in the least rudimentary in function, for they are sensitive, extremely heliotropic, circumnutate at nearly the same rate as the fully developed leaflets and assume when asleep exactly the same position. With *M. pudica* the inner leaflets at the base and between the pinnae are likewise much shortened and obliquely truncated; this fact was well seen in some seedlings of *M. pudica*, in which the third leaf above the cotyledons bore only two pinnae, each with only 3 or 4 pairs of leaflets, of which the inner basal one was less than half as long as its fellow; so that the whole leaf resembled pretty closely that of *M. albida*. In this latter species, the main petiole terminates in a little point, and on each side of this there is a pair of minute, flattened, lancet-shaped projections, hairy on their margins, which drop off and disappear soon after the leaf is fully developed. There can hardly be a doubt that these little projections are the last and fugacious representatives of an additional pair of leaflets to each pinna; for the outer one is twice as broad as the inner one, and a little longer. ...

When the leaves go to sleep, each leaflet twists half round, so as to present its edge to the zenith, and comes into close contact with its fellow. The pinnae also approach each other closely, so that the four terminal leaflets come together. The large basal leaflets (with the little rudimentary ones in contact with them) move inwards and forwards, so as to embrace

Mimosa albida: leaf seen from vertically above.

the outside of the united terminal leaflets, and thus all eight leaflets (the rudimentary ones included) form together a single vertical packet. The two pinnae at the same time that they approach each other sink downwards, and thus instead of extending horizontally in the same line with the main petiole, as during the day, they depend at night at about 45°, or even at a greater angle, beneath the horizon. ...

The torsion or rotation of leaves and leaflets, which occurs in so many cases, apparently always serves to bring their upper surfaces into close approximation with one another, or with other parts of the plant, for their mutual protection. We see this best in such cases as those of *Arachis*, *Mimosa albida*, and *Marsilea*, in which all the leaflets form together at night a single vertical packet. If with *Mimosa pudica* the opposite leaflets had merely moved upwards, their upper surfaces would have come into contact and been well protected; but as it is, they all successively move towards the apex of the leaf; and thus not only their upper surfaces are protected, but the successive pairs become imbricated and mutually protect one another as well as the petioles. This imbrication of the leaflets of sleeping plants is a common phenomenon. ...

Any one who had never observed continuously a sleeping plant, would naturally suppose that the leaves moved only in the evening when going to sleep, and in the morning when awaking; but he would be quite mistaken, for we have found no exception to the rule that leaves which sleep continue to move during the whole twenty-four hours; they move, however, more quickly when going to sleep and when awaking than at other times.

Mitchella repens. Drawing by T. Boys for Conrad Loddiges, *The Botanical Cabinet,*
Consisting of Colored Delineations of Plants.

Mitchella repens

PARTRIDGEBERRY

......

RUBIACEAE—COFFEE FAMILY

FORMS OF FLOWERS, POLLINATION

When it comes to heterostyly, where flower morphs differing in relative stamen and pistil length are produced in the same species (termed pin and thrum morphs), the coffee family is the champ, with more heterostylous genera around the world than any other plant family. Most are tropical—Darwin worked on seventeen genera in the family, having received plants, seeds, and occasionally dried pressed flowers from friends and colleagues far and wide. Patridgeberry, *Mitchella repens*, was the only temperate plant he worked on in the family. An evergreen creeper of eastern North American forests, this species was named by Carl Linnaeus in honor of eighteenth-century Virginia botanist and physician John Mitchell, who sent to his European correspondents seeds and specimens of a great many of the "unknown Beautifull, curious and usefull plants, our Country affords."[98]

Asa Gray, responding to Darwin's queries for examples of dimorphic flowers, brought partridgeberry to his attention. "This would be a good plant for you to experiment upon," he wrote, pointing out its fully dioecious relatives, with their separate-sex flowers.[99] Darwin was sure that such floral dimorphisms, which he first noticed with *Primula* (see p. 271), represented a step in the evolution of separate sexes in flowering plants, and he performed crossing experiments to determine the inter-fertility of different morphs.

Gray sent live partridgeberry specimens in 1862, and when they arrived, Darwin enthused that they "looked as fresh as if dug up the day before! What a pretty little creeper it is with scarlet berries."[100] The fruit, actually a drupe, results from the fusion of the ovaries of the paired trumpet-shaped flowers. In experiments conducted over two seasons, Darwin confirmed that between-morph crosses had a far higher success rate than those within the same morph. But he also noted that partridgeberry was found to occur in functionally dioecious forms as well, with imperfectly formed pistils in one and rudimentary stamens in the other—de facto female and male flowers, even though they technically had the reproductive structures of both sexes. Citing the observations of Thomas Meehan,[101] a British-born botanist who started a nursery in the United States, Darwin declared that "Should these statements be confirmed, *Mitchella* will be proved to be heterostyled in one district and dioecious in another," and he argued in *Forms of Flowers* that *Mitchella* and relatives in the Rubiaceae family thus provided evidence that dioecious species "owe their origin to the transformation of heterostyled plants."[102]

In a twist that would have intrigued Darwin, twentieth-century botanists developed a technique for identifying "functional" gender based on the relative seed production of pin versus thrum morphs. In its native habitat, partridgeberry populations vary in functional gender, and these do not necessarily correspond to apparent gender based on morph—stamen-dominated thrums in some cases produce more seeds than pistil-dominated pins.[103] Modern thinking is thus partially in agreement with Darwin—heterostyly is an adaptation to promote outcrossing—but does not hold that the morphs necessarily represent an evolution toward dioecy.

From: *The Different Forms of Flowers on Plants of the Same Species* (1877)

Mitchella repens.—Prof. Asa Gray sent me several living plants collected when out of flower, and nearly half of these proved long-styled, and the other half short-styled. The white flowers, which are fragrant and which secrete plenty of nectar, always grow in pairs with their ovaries united, so that the two together produce "a berry-like double drupe." In my first series of experiments (1864), I did not suppose that this curious arrangement of the flowers would have any influence on their fertility; and in several instances, only one of the two flowers in a pair was fertilised; and a large proportion or all of these failed to produce berries. In the ensuing year, both flowers of each pair were invariably fertilised in the same manner; and the latter experiments alone serve to show the proportion of flowers which yield berries, when legitimately and illegitimately fertilised; but for calculating the average number of seeds per berry I have used those produced during both seasons.

In the long-styled flowers, the stigma projects just above the bearded throat of the corolla, and the anthers are seated some way down the tube. In the short-styled flowers, these organs occupy reversed positions. In this latter form, the fresh pollen-grains are a little larger and more opaque than those of the long-styled form. ...

... 88 per cent of the paired flowers of both forms, when legitimately fertilised, yielded double berries, nineteen of which contained on an average 4.4 seeds, with a maximum in one of 8 seeds. Of the illegitimately fertilised paired flowers, only 18 per cent yielded berries, six of which contained on an average only 2.1 seeds, with a maximum in one of 4 seeds. Thus the two legitimate unions are more fertile than the two illegitimate, according to the proportion of flowers which yielded berries, in the ratio of 100 to 20; and according to the average number of contained seeds as 100 to 47.

Three long-styled and three short-styled plants were protected under separate nets, and they produced altogether only 8 berries, containing on an average only 1.5 seed. Some additional berries were produced which contained no seeds. The plants thus treated were therefore excessively sterile, and their slight degree of fertility may be attributed in part to the action of the many individuals of Thrips which haunted the flowers. ...

... In the 'Genera Plantarum,' by Bentham and Hooker, the Rubiaceae are divided into 25 tribes, containing 337 genera; and it deserves notice that the genera now known to be heterostyled are not grouped in one or two of these tribes, but are distributed in no less than eight of them. From this fact we may infer that most of the genera have acquired their heterostyled structure independently of one another; that is, they have not inherited this structure from some one or even two or three progenitors in common.

The Bee-flower.

Orchis apifera. Water and bodycolor on vellum by English School artist,
Album of Garden Flowers.

ORCHIDS, FORMS OF FLOWERS, POLLINATION,
CROSS AND SELF-FERTILIZATION

Ophrys orchids, a genus with well over 100 terrestrial species, are the premier example of sexual deception in plants. These orchids deploy visual as well as chemical mimicry (sex pheromones) to entice male insects to attempt mating with the flower, termed pseudo-copulation. What the duped males get for their efforts is a packet of pollinia glued to their body, without the reward of either nectar or sex.104 The common names of two widespread European species, the fly orchid (*Ophrys insectifera*) and the bee orchid (*O. apifera*), reflect the insects they attract. Darwin found *Ophrys apifera* while vacationing at the seaside town of Torquay, and it presented a special puzzle—a tendency to self-fertilize in Britain, in the northern part of its range, with few insect visitors arriving in spite of its apparent "efforts" at attraction. His letters to colleagues noted his concerns. "No single point in natural history interests and perplexes me as much as the self-fertilization of the bee orchis,"105 he wrote to one correspondent, and to another he lamented that "all facts point clearly to eternal self-fertilisation; yet I cannot swallow the bitter pill" that *Ophrys* is self-fertilized.106 After an attempt at crowd-sourcing, with a letter published in the *Gardeners' Chronicle* that presented his observations and encouraged readers to study the fertilization of the plants in locales where insects might be more abundant, Darwin concluded that, to avoid extinction, plants such as the bee orchid are capable of producing viable seeds by self-fertilisation in order to survive times when insect pollinators are scarce.

From: "Fertilisation of British orchids by insect agency." *Gardeners' Chronicle and Agricultural Gazette* (June 1860, p. 528)

I should be extremely much obliged to any person living where the Bee or Fly Orchis is tolerably common, if he will have the kindness to make a few simple observations on their manner of fertilisation. ... [In] the Bee Orchis (*Ophrys apifera*) ... the pollen masses are furnished with sticky glands, but differently from in [other] Orchids, they naturally fall out of their pouches; and from being of the proper length, though still retained at the gland-end, they fall on the stigmatic surface, and the plant is thus self-fertilised. During several years I have examined many flowers, and never in a single instance found even one of the pollen-masses carried away by insects, or ever saw the flower's own pollen-masses fail to fall on the stigma. Robert Brown consequently believed that the visits of insects would be injurious to the fertilisation of this Orchis; and rather fancifully imagined that the flower resembled a bee in order to deter their visits. We must admit that the natural falling out of the pollen-masses of this Orchis is a special contrivance for its self-fertilisation; and as far as my experience goes, a perfectly successful contrivance, for I have always found this plant self-fertilised; nevertheless a long course of observation has made me greatly doubt whether the flowers of any kind of plant are for a perpetuity of generations fertilised by their own pollen. And what are we to say with respect to the sticky glands of the Bee Orchis, the use and efficiency of which glands in all other British Orchids are so manifest? Are we to conclude that this one species is provided with these organs for no use? I cannot think so; but would rather infer that, during some years or in some other districts, insects do visit the Bee Orchis and occasionally transport pollen from one flower to another, and thus give it the advantage of an occasional cross. So with the Bee Orchis, though its self-fertilisation is specially provided for, it may not exist here under the most favourable conditions of life; and in other districts or during particular seasons it may be visited by insects, and in this case, as its pollen masses are furnished with sticky glands, it would almost certainly receive the benefit of an occasional cross impregnation. It

is this curious apparent contradiction in the structure of the Bee Orchis—
one part, namely the sticky glands, being adapted for fertilisation by insect
agency—another part, namely the natural falling out of the pollen-masses,
being adapted for self-fertilisation without insect agency—which makes me
anxious to hear what happens to the pollen-masses of the Bee Orchis in other
districts or parts of England.

From: *The Various Contrivances by Which
Orchids are Fertilised by Insects* (2nd ed., 1877)

The Bee Ophrys differs widely from the great majority of Orchids in being
excellently constructed for fertilising itself. The two pouch-formed ros-
tella, the viscid discs, and the position of the stigma, are nearly the same
as in the other species of *Ophrys*; but the distance of the two pouches
from each other, and the shape of the pollen-masses are somewhat variable. The caudicles of the pollinia are remarkably long, thin, and flexible,
instead of being, as in all the other Ophreae seen by me, rigid enough to
stand upright. They are necessarily curved forward at their upper ends,
owing to the shape of the anther-cells; and the pear-shaped pollen-masses
lie embedded high above and directly over the stigma. The anther-cells
naturally open soon after the flower is fully expanded, and the thick ends
of the pollen-masses then fall out, the viscid discs still remaining in their
pouches. Slight as is the weight of the pollen-masses, yet the caudicles are
so thin and quickly become so flexible, that in the course of a few hours
they sink down, until they hang freely in the air (see lower pollen-mass
in fig. A) exactly opposite to and in front of the stigmatic surface. In this
position a breath of air, acting on the expanded petals, sets the flexible and
elastic caudicles vibrating, and they almost immediately strike the viscid
stigma, and, being there secured, impregnation is effected. To make sure
that no other aid was requisite, though the experiment was superfluous, I
covered up a plant under a net, so that the wind, but no insects, could pass
in, and in a few days the pollinia became attached to the stigmas. But the

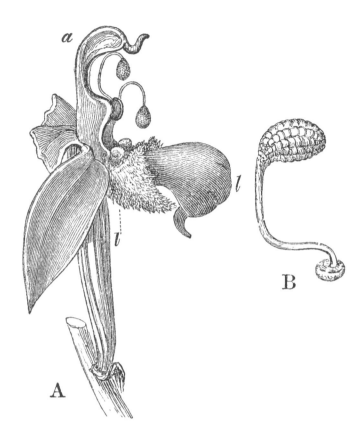

Ophrys apifera, or Bee Ophrys. A. Side view of flower, with the
upper sepal and the two upper petals removed. One pollinium,
with its disc still in its pouch, is represented as just falling out
of the anther-cell; and the other has fallen almost to its full
extent, opposite to the hidden stigmatic surface. B. Pollinium in
the position in which it lies embedded. *a.* anther *l.* labellum

pollinia of a spike kept in water in a still room remained free, suspended in front of the stigma, until the flowers withered. ...

I have often noticed that the spikes of the Bee Ophrys apparently produced as many seed-capsules as flowers; and near Torquay I carefully examined many dozen plants, some time after the flowering season; and on all I found from one to four, and occasionally five, fine capsules, that is, as many capsules as there had been flowers. In extremely few cases, with the exception of a few deformities, generally on the summit of the spike, could a flower be found which had not produced a capsule. Let it be observed what a contrast this species presents with the Fly Ophrys, which requires insect aid for its fertilisation, and which from forty-nine flowers produced only seven capsules! ...

That cross-fertilisation is beneficial to most Orchids we may infer from the innumerable structures serving for this purpose which they present; and I have elsewhere shown in the case of many other groups of plants that the benefits thus derived are of high importance. On the other hand, self-fertilisation is mainfestly advantageous in as far as it ensures a full supply of seed; and we have seen with the other British species of *Ophrys* which cannot fertilise themselves, how small a proportion of their flowers produce capsules. Judging therefore from the structure of the flowers of *O. apifera*, it seems almost certain that at some former period they were adapted for cross-fertilisation, but that failing to produce a sufficiency of seed they became slightly modified so as to fertilise themselves. ...

It deserves especial attention that the flowers of all the above-named self-fertile species still retain various structures which it is impossible to doubt are adapted for insuring cross-fertilisation, though they are now rarely or never brought into play. We may therefore conclude that all these plants are descended from species or varieties which were formerly fertilised by insect-aid. Moreover, several of the genera to which these self-fertile species belong, include other species, which are incapable of self-fertilisation.

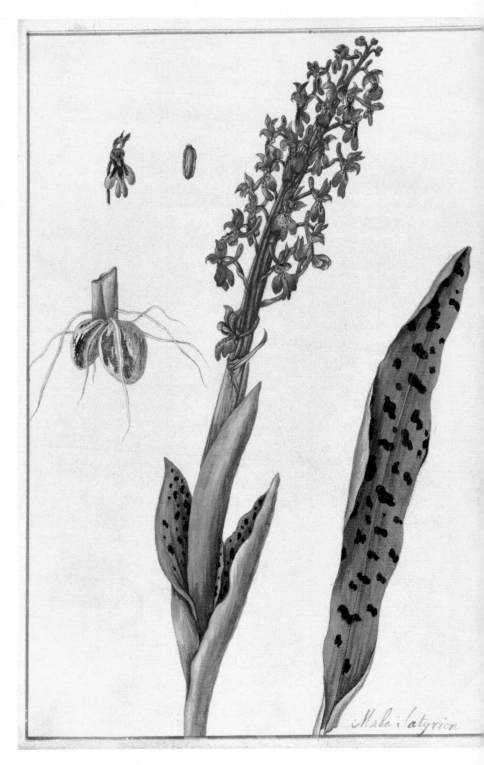

Orchis mascula. Watercolor by Elizabeth Blackwell, *A Curious Herbal.*

Orchis

ORCHID

———— ...... ————

ORCHIDACEAE—ORCHID FAMILY

ORCHIDS, FORMS OF FLOWERS, POLLINATION

It may seem odd that Darwin followed up on his epochal work *On the Origin of Species* with a book on orchids, a subject seemingly far narrower in scope. But he saw great significance in the intricacies of orchid flowers, declaring that in studying orchids, "hardly any fact has struck me so much as the endless diversity of structure,—the prodigality of resources,—for gaining the very same end, namely, the fertilisation of one flower by pollen from another plant. This fact is to a large extent intelligible on the principle of natural selection."[107] His book thus had an important subtext: "it was a 'flank movement' on the enemy," he confessed to Asa Gray, "It bears on design—that endless question."[108]

The name "orchid" is derived from the Ancient Greek term for testicles, in reference to the paired ovate root tubers so common in the group. Carl Linnaeus published the first monograph on orchids in 1740, a work that served as the basis of his landmark *Species Plantarum* of 1753, where the type genus *Orchis* was named and a host of orchid species were first given binomials. Following Linnaeus's tendency to be more "lumper" than "splitter," early on, just about all orchids were considered genus *Orchis*, but later systematic and phylogenetic analyses led to the naming of a host of additional genera. Today about twenty-one *Orchis* species are recognized, all Old World and ranging from Europe and Eurasia to north Africa. Several native orchids grew near Darwin's home at a favorite site known to him and his family as

Orchis Bank, now preserved by the Kent Wildlife Trust. *Orchis mascula* was common there, as was *Orchis* (=*Anacamptis*) *pyramidalis*.

The structure of *Orchis* flowers served as a reference for Darwin to compare and relate pollination mechanisms of other orchid genera. *Orchis* species were among his favorites. "I declare I think its adaptations in every part of flower quite as beautiful and plain, or even more beautiful, than in Woodpecker," he wrote to Joseph Hooker, at Kew, "I never saw anything so beautiful."[109] He was especially enthralled with the lock-and-key relationship between orchids and their pollinators, and the various "contrivances" by which the pollen packets (pollinia) with their sticky bases are glued to the insects. In one study, he monitored insects actively pollinating some flowers, covered others with a bell jar to prevent visitation as test cases, and, in another, simulated the visit of a pollinator by using a pencil tip to extract the pollinia, marveling how the stalks bend over within seconds of being removed to position themselves for delivering pollen to the next flower. He first noticed this in 1860, sketching it out in a letter to naturalist (and fellow orchid enthusiast) Alexander More, and describing how the stigmas of this species are held laterally to receive the pollinia: "Is this not a pretty relation to visits of insects?"[110]

[**From: *The Various Contrivances by Which Orchids are Fertilised by Insects* (2nd ed., 1877)**]

Orchis mascula—The accompanying diagrams show the relative position of the more important organs in the flower of the Early Orchis. The sepals and the petals have been removed, excepting the labellum with its nectary. The nectary is shown only in the side view (*n*, fig. A); for its enlarged orifice is almost hidden in shade in the front view (B). The stigma (*s*) is bilobed and consists of two almost confluent stigmas; it lies under the pouch-formed rostellum (*r*). The anther (*a*, in B and A) consists of two rather widely separated cells, which are longitudinally open in front: each cell includes a pollen-mass or pollinium.

A pollinium removed out of one of the two anther-cells is represented by fig. C; it consists of a number of wedge-formed packets of pollen-grains (see fig. F, in which the packets are forcibly separated), united together by excessively elastic, thin threads. These threads become confluent at the lower end of each pollen-mass, and compose the straight elastic caudicle

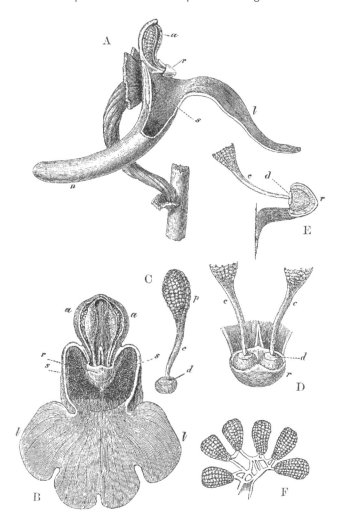

Orchis mascula. a. anther, consisting of two cells. *n.* nectary. *r.* rostellum. *p.* pollen-mass. *s.* stigma. *c.* caudicle of pollinium. *l.* labellum. *d.* viscid disc of pollinium.

(*c*, C). The end of the caudicle is firmly attached to the viscid disc (*d*, C), which consists (as may be seen in the section of the pouch-formed rostellum, fig. E) of a minute oval piece of membrane, with a ball of viscid matter on its underside. Each pollinium has its separate disc; and the two balls of viscid matter lie enclosed together (fig. D) within the rostellum.

Now let us see in the case of *Orchis mascula* how this complex mechanism acts. Suppose an insect to alight on the labellum, which forms a good landing-place, and to push its head into the chamber (see side view, A, or front view, B), at the back of which lies the stigma (s), in order to reach with its proboscis the end of the nectary; or, which does equally well to show the action, push very gently a sharply-pointed common pencil into the nectary. Owing to the pouch-formed rostellum projecting into the gangway of the nectary, it is scarcely possible that any object can be pushed into it without the rostellum being touched. The exterior membrane of the rostellum then ruptures in the proper lines, and the lip or pouch is easily depressed. When this is effected, one or both of the viscid balls will almost infallibly touch the intruding body. So viscid are these balls that whatever they touch they firmly stick to. Moreover, the viscid matter has the peculiar chemical quality of setting, like a cement, hard and dry in a few minutes' time. As the anther-cells are open in front, when the insect withdraws its head, or when the pencil is withdrawn, one pollinium, or both, will be withdrawn, firmly cemented to the object, projecting up like horns, as shown ... by the upper figure, A. The firmness of the attachment of the cement is very necessary, for if the pollinia were to fall sideways or backwards they could never fertilise the flower. From the position in which the two pollinia lie in their cells, they diverge a little when attached to any object. Now suppose that the insect flies to another flower, or let us insert the pencil (A), with the attached pollinium, into the same or into another nectary; by looking at the diagram ... it will be evident that the firmly attached pollinium will be simply pushed against or into its old position, namely, into the anther-cell. How then can the flower be fertilised? This is effected by a beautiful contrivance: though the viscid surface remains immovably affixed, the apparently insignificant and minute disc of membrane to which the caudicle adheres is endowed with a remarkable power of contraction ... which causes the pollinium to sweep through an angle of about ninety degrees, always in one direction, viz., towards the apex of the proboscis or pencil, in the course of thirty seconds on an average. The position of the pollinium after the movement is shown at B. After this movement, completed in an interval of time which would allow an insect to fly to another plant, it will be seen, by turning to the diagram ... that, if the pencil be inserted into the nectary, the thick end of the pollinium now exactly strikes the stigmatic surface.

I have now described the structure of most of the British and of a few foreign species in the genus *Orchis* and its close allies. All these species require the aid of insects for their fertilisation. This is obvious from the fact that the pollinia are so closely embedded in the anther-cells, and the ball of viscid matter in the pouch-formed rostellum, that they cannot be shaken out by violence. We have also seen that the pollinia do not assume the proper position for striking the stigmatic surface until some time has elapsed; and this indicates that they are adapted to fertilise, not their own flowers,

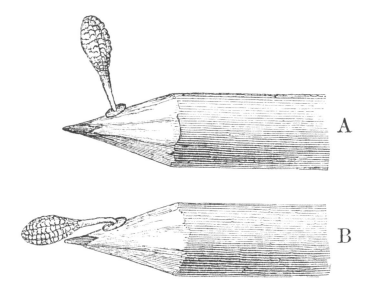

A. Pollen-mass of *O. mascula*, when first attached. B. Pollen-mass of
O. mascula, after the act of depression.

but those on a distinct plant. To prove that insects are necessary for the fertilisation of the flowers, I covered up a plant of *Orchis morio* under a bell-glass, before any of its pollinia had been removed, leaving three adjoining plants uncovered; I looked at the latter every morning, and daily found some of the pollinia removed, till all were gone with the exception of those in a single flower low down on one spike, and of those in one or two flowers on the summits of all the spikes, which were never removed. But it should be observed that when only a very few flowers remain open on the summits of the spikes, these are no longer conspicuous, and would consequently be rarely visited by insects. I then looked at the perfectly healthy plant under the bell-glass, and it had, of course, all its pollinia in the anther-cells. I tried an analogous experiment with specimens of O. mascula with the same result. It deserves notice that the spikes which had been covered up, when subsequently left uncovered, never had their pollinia carried away by insects, and did not, of course, set any seed, whereas the adjoining plants produced plenty of seed. From this fact it may be inferred that there is a proper season for each kind of *Orchis*, and that insects cease their visits after the proper season has passed. ...

I have examined spikes of O. pyramidalis in which every single expanded flower had its pollinia removed. The forty-nine lower flowers of a spike from Folkestone (sent me by Sir Charles Lyell) actually produced forty-eight fine seed-capsules; and of the sixty-nine lower flowers in three other spikes, seven alone had failed to produce capsules. These facts show how well moths and butterflies perform their office of marriage-priests. ...

It has been shown how numerous and beautiful are the contrivances for the fertilisation of Orchids. We know that it is of the highest importance that the pollinia, when attached to the head or proboscis of an insect, should be fixed symmetrically, so as not to fall either sideways or backwards. We know that in the species as yet described the viscid matter of the disc sets hard in a few minutes when exposed to the air, so that it would be a great advantage to the plant if insects were delayed in sucking the nectar, time being thus allowed for the disc to become immovably affixed.

Oxalis acetosella. Water and bodycolor on vellum by English School artist,
Album of Garden Flowers.

Oxalis
WOOD SORREL

———— ...... ————

OXALIDACEAE—WOOD SORREL FAMILY

FORMS OF FLOWERS, PLANT MOVEMENT

Wood sorrels, shamrocks, and other species in the cosmopolitan genus *Oxalis* have features that led Darwin to study the plants on several levels, from heterostyly to leaf and fruit movement. In the early 1860s, when Darwin first became aware that the flowers of some species exhibit morphs with varied lengths of pistils and stamens (see p. 271), he queried his botanist friends for more cases. He surmised that the different morphs were erroneously interpreted as distinct species, something that the distinguished English botanist George Bentham confirmed in sending Darwin a list of *Oxalis* species varying in stamens and pistil length, many of which he and others suspected were actually morphs of each other.[111]

Ultimately, with the help of South African naturalist Roland Trimen and German botanist Friedrich Hildebrand, he was able to present data for thirteen *Oxalis* species from around the world in *Forms of Flowers*. Their work focused on trimorphic flowers that produce three distinct stamen and pistil length morphs, which practically cried out for experiments testing "legitimate" and "illegitimate" crosses. Darwin also pondered unusual "homostyled" *Oxalis* species such as the common European wood sorrel *O. acetosella*, with single-morph flowers, and specialized cleistogamous flowers that set seed in abundance despite never opening in bloom.

From: *The Different Forms of Flowers on Plants of the Same Species* (1877)

In 1863 Mr. Roland Trimen wrote to me from the Cape of Good Hope that he had there found species of *Oxalis* which presented three forms; and of these he enclosed drawings and dried specimens. Of one species he collected 43 flowers from distinct plants, and they consisted of 10 long-styled, 12 mid-styled, and 21 short-styled. Of another species he collected 13 flowers, consisting of 3 long-styled, 7 mid-styled, and 3 short-styled. In 1866 Prof. Hildebrand proved by an examination of the specimens in several herbaria that 20 species are certainly heterostyled and trimorphic, and 51 others almost certainly so. He also made some interesting observations on living plants belonging to one form alone; for at that time he did not possess the three forms of any living species. During the years 1864 to 1868, I occasionally experimented on *Oxalis speciosa*, but until now have never found time to publish the results. ... I may premise that in all the species seen by me, the stigmas of the five straight pistils of the long-styled form stand on a level with the anthers of the longest stamens in the two other forms. In the mid-styled form, the stigmas pass out between the filaments of the longest stamens and they stand rather nearer to the upper anthers than to the lower ones. In the short-styled form, the stigmas also pass out between the filaments nearly on a level with the tips of the sepals. The anthers in this latter form and in the mid-styled rise to the same height as the corresponding stigmas in the other two forms. ...

Oxalis speciosa.*—This species, which bears pink flowers, was introduced from the Cape of Good Hope. A sketch of the reproductive organs of the three forms [is shown]. The stigma of the long-styled form (with the papillae on its surface included) is twice as large as that of the short-styled, and that of the mid-styled intermediate in size. ...

---

* Now *Oxalis variabilis*, a South African species.

... Thirty-six flowers on the three forms legitimately fertilised yielded 30 capsules, these containing on an average 58.36 seeds. Ninety-five flowers illegitimately fertilised yielded 12 capsules, containing on an average 28.58 seeds. Therefore, the fertility of the six legitimate to that of the twelve illegitimate unions, as judged by the proportion of flowers which yielded capsules, is as 100 to 15, and judged by the average number of seeds per capsule as 100 to 49. ...

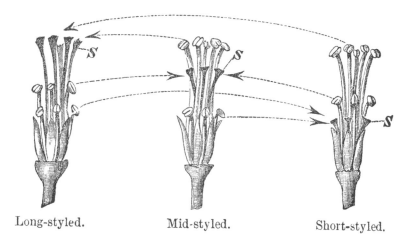

Long-styled.   Mid-styled.   Short-styled.

Oxalis speciosa (with the petals removed). *S*, stigmas. The dotted lines with arrows show which pollen must be carried to the stigmas for legitimate fertilisation.

Darwin's later research, conducted with his son Francis, involved various kinds of movement of stems, leaves, flowers, and fruit capsules. Nighttime folding of leaves, for example (for which they coined the term "nyctitropism"), was widely known, but there was no accepted explanation at the time. Linnaeus and others hinted at a protective function, but Darwin and son approached the question experimentally, testing several species including *Oxalis acetosella* and the Chilean *O. carnosa*, which fold their three leaves like a closed umbrella at night.

In one series of experiments, they pinned or tied leaves open and exposed the plants to subfreezing temperatures outdoors to determine if folding protects from radiative heat loss and frost damage. It does. "I think we have *proved* that the sleep of plants is to lessen injury to leaves from radiation," he wrote to Joseph Hooker at the Royal Botanic Gardens, Kew, "this has interested me much and has cost us great labour, as it has been a problem since the time of Linnaeus." The revelation was not without cost, however, and he lamented to Hooker, "we have killed or badly injured a multitude of plants."[112]

In other studies of leaves, shoots, and fruits, Darwin tracked movements using tracing paper on glass plates, marking patterns of growth and slow circumnutation, circular movements, hour by hour and day by day. He noted and defined movements such as "apogeotropism" (growth away from the earth), "hyponasty" (growth along a lower surface causing leaves and stems to bend upward), and "epinasty" (increased growth on the upper surface, causing downward bending).

[From: *The Power of Movement in Plants* (1880)]

We will now describe in detail the experiments which were tried. These were troublesome from our not being able to predict how much cold the leaves of the several species could endure. Many plants had every leaf killed, both those which were secured in a horizontal position and those which were allowed to sleep—that is, to rise up or sink down vertically. Others again had not a single leaf in the least injured, and these had to be re-exposed either for a longer time or to a lower temperature.

Oxalis acetosella.—A very large pot, thickly covered with between 300 and 400 leaves, had been kept all winter in the greenhouse. Seven leaves were pinned horizontally open and were exposed on March 16th for 2 h. to a clear sky, the temperature on the surrounding grass being -4° C. (24° to 25° F.). Next morning all seven leaves were found quite killed, so were many of the free ones which had previously gone to sleep, and about 100 of them, either dead or browned and injured were picked off. Some leaves showed that they had been slightly injured by not expanding during the whole of the next day, though they afterwards recovered. As all the leaves which were pinned open were killed, and only about a third or fourth of the others were either killed or injured, we had some little evidence that those which were prevented from assuming their vertically dependent position suffered most.

The following night (17th) was clear and almost equally cold (-3° to -4° C. on the grass), and the pot was again exposed, but this time for only 30 m. Eight leaves had been pinned out, and in the morning two of them were dead, whilst not a single other leaf on the many plants was even injured.

Considering these cases, there can be no doubt that the leaves of this *Oxalis*, when allowed to assume their normal vertically dependent position at night, suffer much less from frost than those which had their upper surfaces exposed to the zenith. ...

The difference in the amount of dew on the pinned-open leaflets and on those which had gone to sleep was generally conspicuous; the latter being sometimes absolutely dry, whilst the leaflets which had been horizontal were coated with large beads of dew. This shows how much cooler the leaflets fully exposed to the zenith must have become, than those which stood almost vertically, either upwards or downwards, during the night.

From the several cases above given, there can be no doubt that the position of the leaves at night affects their temperature through radiation to such a degree, that when exposed to a clear sky during a frost, it is a question of life and death. We may therefore admit as highly probable, seeing that their nocturnal position is so well adapted to lessen radiation, that the object gained by their often complicated sleep movements, is to lessen the degree to which they are chilled at night. It should be kept in mind that it is especially the upper surface which is thus protected, as it is never directed towards the zenith, and is often brought into close contact with the upper surface of an opposite leaf or leaflet. ...

In most of the species in this large genus the three leaflets sink vertically down at night; but as their sub-petioles are short, the blades could not assume this position from the want of space, unless they were in some manner rendered narrower; and this is effected by their becoming more or less folded (Figure below).

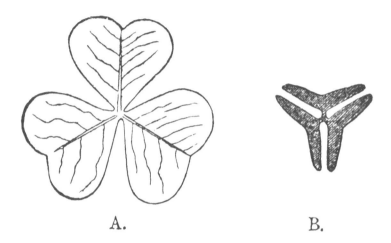

A.   B.

Oxalis acetosella: A, leaf seen from vertically above; B, diagram of leaf asleep, also seen from vertically above.

Passiflora caerulea. Bodycolor on vellum by Jan Withoos, *Dutch Florilegium.*

Passiflora

PASSION FLOWER

......

PASSIFLORACEAE—PASSION FLOWER FAMILY

CLIMBING PLANTS

Passion flowers, introduced to Europe from the New World in the sixteenth century, were hugely popular in Darwin's era, cultivated in botanical gardens and private greenhouses where their showy, intricate, and colorful flowers captivated plant lovers and led to cultivation of countless varieties. There are some 550 species in the genus *Passiflora* worldwide, mainly from Central and South America, and most of them herbaceous climbers. Darwin's interest in this group began with questions about hybridization and their inability to self-pollinate—a mystery at the time—and soon expanded to include pollinators and seed dispersal. Later, in the 1860s, he turned his attention to the climbing abilities of *Passiflora*, and there his interest turned into a passion.

The tendrils intrigued him, as they appeared to be unusual in their development. It might be natural to assume that the tendrils of different climbing plant groups are all derived from basically the same ancestral structure—leaves, for example—but Darwin became convinced that different parts became modified in the same way in different groups, converging on the tendril morphology. His inference grew out of comparative studies facilitated by Joseph Hooker at the Royal Botanic Gardens, Kew: "I am getting very much amused by my tendrils," he wrote to Hooker. "Will you just think whether you know any plant, which you could give or lend me or I could buy with tendrils remarkable in any way, for development, for odd or peculiar structure or even for odd place in natural arrangement ... It is really curious the diversification of irritability."[113] That "diversification" is key—cases of diversified structures evolving into the same strikingly convergent structure and function spoke to him of the power of selection.

Darwin worked on four species of passion flowers. He asked his son William to make detailed microscopic observations of their tendril development and concluded from his sketches that, in this group, tendrils are derived from peduncles, or flower stems.[114] He was also struck by the rapidity of the circular searching movement (circumnutation) of some *Passiflora* tendrils and their exquisite sensitivity to touch. *Passiflora gracilis*, the champion on both counts, was further intriguing because the internodes of young shoots rotated to boot. Darwin thought that might be a sign of a shoot-twining past. Do any living passion flowers or their relatives have twining shoots, he wondered? He could find none, so he asked Daniel Oliver, keeper of the Herbarium at Kew, to put the question to his fellow botanists: "At any time when botanists congregate thickly, or you come across any one who has studied the order of Passiflorae, will you ask the assembly whether any member of the order climbs without the aid of tendrils i.e. spirally twines; I am really anxious to know."[115] This was very much in keeping with Darwin's crowd-sourcing, and his enthusiasm and curiosity were infectious. "I have at last got a flower of *Passiflora Princeps* [now *P. racemosa*]," wrote another correspondent, Thomas Farrer, after being exhorted to make observations. "You were quite right, as you seem always to be. ... We have been deeply interested in watching the wonderful motions of Passifloras in climbing. They seek and find and hold on and pull up like an animal."[116] Yes, Darwin surely thought, that was just it!

From: *The Movements and Habits of Climbing Plants*
(2nd ed., 1875)

Passiflora gracilis.—This well-named, elegant, annual species differs from the other members of the group observed by me, in the young internodes having the power of revolving. It exceeds all the other climbing plants which I have examined, in the rapidity of its movements, and all tendril-bearers in the sensitiveness of the tendrils. The internode which carries the upper active tendril and which likewise carries one or two younger immature internodes, made three revolutions, following the sun, at an average rate of 1 hr. 4 m.; it then made, the day becoming very hot, three other revolutions at an average rate of between 57 and 58 m.; so that the average of all six revolutions was 1 hr. 1 m. The apex of the tendril describes elongated ellipses, sometimes narrow and sometimes broad, with their longer axes inclined in slightly different directions. The plant can ascend a thin upright stick by the aid of its tendrils; but the stem is too stiff for it to twine spirally round it, even when not interfered with by the tendrils, these having been successively pinched off at an early age.

When the stem is secured, the tendrils are seen to revolve in nearly the same manner and at the same rate as the internodes. The tendrils are very thin, delicate, and straight, with the exception of the tips, which are a little curved; they are from 7 to 9 inches in length. A half-grown tendril is not sensitive; but when nearly full-grown they are extremely sensitive. A single delicate touch on the concave surface of the tip soon caused one to curve; and in 2 minutes it formed an open helix. A loop of soft thread weighing $\frac{1}{32}$nd of a grain (2.02 mg.) placed most gently on the tip thrice caused distinct curvature. A bent bit of thin platina wire weighing only $\frac{1}{50}$th of a grain (1.23 mg.) twice produced the same effect; but this latter weight, when left suspended, did not suffice to cause a permanent curvature. These trials were made under a bell-glass, so that the loops of thread and wire were not agitated by the wind. The movement after a touch is very rapid: I took hold of the lower part of several tendrils, and then touched their concave tips with a thin twig and watched them carefully through a lens; the tips evidently began to bend after the following intervals—31, 25, 32, 31, 28, 39, 31, and 30 seconds; so that the movement was generally perceptible in half a minute after a touch; but on one occasion it was distinctly visible in 25 seconds. One of the tendrils which thus became bent in 31 seconds, had been touched two hours previously and had coiled into a helix; so that in this interval it had straightened itself and had perfectly recovered its irritability. ...

Passiflora quadrangularis.—This is a very distinct species. The tendrils are thick, long, and stiff; they are sensitive to a touch only on the concave surface towards the extremity. When a stick was placed so that the middle of the tendril came into contact with it, no curvature ensued. In the hothouse, a tendril made two revolutions, each in 2 hrs. 22 m.; in a cool room, one was completed in 3 hrs., and a second in 4 hrs. The internodes do not revolve; nor do those of the hybrid *P. floribunda.* ...

The tendrils of many kinds of plants, if they catch nothing, contract after an interval of several days or weeks into a spire; but in these cases, the movement takes place after the tendril has lost its revolving power and hangs down; it has also then partly or wholly lost its sensibility; so that this movement can be of no use. The spiral contraction of unattached tendrils is a much slower process than that of attached ones. Young tendrils which have caught a support and are spirally contracted, may constantly be seen on the same stem with the much older unattached and uncontracted tendrils. ... A full-grown tendril of *Passiflora quadrangularis* which had caught a stick began in 8 hrs. to contract, and in 24 hrs. formed several spires; a younger tendril, only two-thirds grown, showed the first trace of contraction in two days after clasping a stick, and in two more days formed several spires. It appears, therefore, that the contraction does not begin until the tendril is grown to nearly its full length. Another young tendril of about the same age and length as the last did not catch any object; it acquired its full length in four days; in six additional days it first became flexuous, and in two more days formed one complete spire. This first spire was formed towards the basal end, and the contraction steadily but slowly progressed towards the apex; but the whole was not closely wound up into a spire until

21 days had elapsed from the first observation, that is, until 17 days after the tendril had grown to its full length. ...

The spiral contraction which ensues after a tendril has caught a support is of high service to the plant; hence its almost universal occurrence with species belonging to widely different orders. When a shoot is inclined and its tendril has caught an object above, the spiral contraction drags up the shoot. When the shoot is upright, the growth of the stem, after the tendrils have seized some object above, would leave it slack, were it not for the spiral contraction which draws up the stem as it increases in length. Thus there is no waste of growth, and the stretched stem ascends by the shortest course.

The Scarlet-been.

Phaseolus coccineus. Water and bodycolor on vellum by Dame Ann Hamilton,
Drawings of Plants.

Phaseolus

BEANS

———— ...... ————

FABACEAE—PEA FAMILY

FORMS OF FLOWERS, POLLINATION, CROSS AND SELF-FERTILIZATION

String beans, black beans, pinto beans, scarlet runner beans—all are in the genus *Phaseolus*, with seventy annual and perennial species, five of which are domesticated with varieties grown as bush beans and pole beans, producing green beans and dry beans. The most commonly grown ones—*Phaseolus vulgaris,* common bean, and *P. coccineus*, scarlet runner bean—are both originally from Central America and have been cultivated for millennia.

Darwin surely grew and ate a lot of beans, that staple of the kitchen garden. He first commented on them in connection with pollination in his "Transmutation Notebooks" of the late 1830s and early 1840s, making observations in the gardens at his and his wife Emma's childhood homes in Shrewsbury and nearby Maer Hall. He was fascinated by the curious structure of bean flowers, with stamens and pistil coiled within the tubular keel-petal ("like a French horn"), below which hang two wing-petals. He delighted in how the stamens and pistil are suddenly exserted to dab pollen on unsuspecting large bumblebees (known at the time as "humble-bees") that alight on the wing-petals. He grew very fond of the bees, even rallying to their defense when an irate correspondent in the *Gardeners' Chronicle* called for their eradication, accusing them of damaging the bean crop by perforating the base of the flower to steal the nectar. Darwin wrote a letter in reply, acknowledging that the bees may be guilty of not extracting nectar "in the manner nature intended them" and so indirectly affecting seed set by failing to pollinate. But he urged forgiveness. "Although I can believe that such wicked bees may be injurious to the seedsman," Darwin continued, "one would lament to see these industrious, happy-looking creatures punished with the severity proposed by your correspondent."[117]

He wrote other letters to the *Gardeners' Chronicle* detailing the bees' behavior and the fertilization of beans,[118] weighing in on an ongoing debate over to what extent crossing (by insects) was necessary for bean fertilization, while crowdsourcing by appealing to readers for observations from their own gardens. A Hampshire gardener named Henry Coe even did an experiment for him, planting twelve lots of bean plants each distinctive in seed color, and carefully collecting the beans eventually produced by each. Analyzing samples sent by Coe, Darwin marveled at the variation: "Beans of new colours have appeared, such as pure white, bright purple, yellow, and many are much mottled. Not one of the twelve lots has transmitted its own tint to all the Beans produced by it; nevertheless, the dark Beans have clearly produced a greater number of dark, and the light coloured Beans a greater number of light colour. The mottling seems to have been strongly inherited, but always increased."[119] The explanation to Darwin was clear: though beans self-pollinate pretty well, insects really mix things up.

> ## From: "Bees and the fertilisation of kidney beans." *Gardeners' Chronicle and Agricultural Gazette* (24 October 1857)

Every one who has looked at the flower of the Kidney Bean must have noticed in how curious a manner the pistil with its tubular keel-pistil curls like a French horn to the left side—the flower being viewed in front. Bees, owing to the greater ease with which they can reach the copious nectar from the left side, invariably stand on the left wing-petal; their weight and the effort of sucking depresses this petal, which, for its attachment to the

keel-petal, causes the pistil to protrude. On the pistil beneath the stigma
there is a brush of fine hairs, which when the pistil is moved backwards and
forwards, sweeps the pollen already shed out of the tubular and curled
keel-petal, and gradually pushes it on to the stigma. I have repeatedly tried
this by gently moving the wing petals of a lately expanded flower. Hence
the movement of the pistil indirectly caused by the bees would appear
to aid in the fertilisation of the flower by its own pollen; but besides this,
pollen from the other flowers of the Kidney Bean sometimes adheres to
the right side of the head and body of the bees, and this can scarcely fail
occasionally to be left on the humid stigma, quite close to which, on the
left side, the bees invariably insert their proboscis. Believing that the brush
on the pistil, its backward and forward curling movement, its protrusion
on the left side, and the constant alighting of the bees on the same side,
were not accidental coincidences, but were connected with, perhaps necessary to, the fertilisation of the flower, I examined the flowers just before
their expansion. The pollen is then already shed; but from its position just
beneath the stigma, and from its coherence, I doubt whether it could get
on the stigma, without some movement of the wing petals; and I further
doubt whether any movement, which the wind might cause, would suffice.
I may add that all which I have here described occurs in a lesser degree with
Lathyrus grandiflorus. To test the agency of the bees, I put on three occasions a few flowers within bottles and under gauze: half of these I left quite
undisturbed; of the other half I daily moved the left wing-petal, exactly as
a bee would have done whilst sucking. Not one of the undisturbed flowers set a pod, whereas the greater number (but not all) of those which I
moved, and which were treated in no other respect differently, set fine
pods with good seeds. I am aware that this little experiment ought to have
been repeated many times; and I may be greatly mistaken, but my belief at
present is, that if every bee in Britain were destroyed, we should not again
see a pod on our Kidney Beans. These facts make me curious to know the
meaning of Mr. Swayne's allusion to the good arising from the artificial
fertilisation of early Beans. I am also astonished that the varieties of the
Kidney Bean can be raised true when grown near each other. I should have

expected that they would have been crossed by the bees bringing pollen from other varieties; and I should be infinitely obliged for any information on this head from any of your correspondents. As I have mentioned bees, a little fact which surprised me may be worth giving:—One day I saw for the first time several large humble-bees visiting my rows of the tall scarlet Kidney Bean; they were not sucking at the mouth of the flower, but cutting holes through the calyx, and thus extracting the nectar. I watched this with some attention, for though it is a common thing in many kinds of flowers to see humble-bees sucking through a hole already made, I have not very often seen them in the act of cutting. As these humble-bees had to cut a hole in almost every flower, it was clear that this was the first day on which they had visited my Kidney Beans. I had previously watched every day for some weeks, and often several times daily, the hive-bees, and had seen them always sucking at the mouth of the flower. And here comes the curious point: the very next day after the humble-bees had cut the holes, every single hive bee, without exception, instead of alighting on the left wing-petal, flew straight to the calyx and sucked through the cut hole; and so they continued to do for many following days. Now how did the hive-bees find out that the holes had been made? Instinct seems to be here out of the question, as the Kidney Bean is an exotic. The holes could scarcely be seen from any point, and not at all from the mouth of the flower, where the hive-bees hitherto had invariably alighted. I doubt whether they were guided by a stronger odour of the nectar escaping through the cut holes; for I have found in the case of the little blue *Lobelia*, which is a prime favourite of the hive-bee, that cutting off the lower striped petals deceived them; they seem to think the mutilated flowers are withered, and they pass them over unnoticed. Hence I am strongly inclined to believe that the hive-bees saw the humble-bees at work, and well understanding what they were at, rationally took immediate advantage of the shorter path thus made to the nectar.

Beans were called into service in a variety of Darwin's other studies. He reported growth experiments with crossed and selfed plants in *Cross and Self Fertilisation*, discussed their twining abilities in *Climbing Plants*, and, in *Movement*, documented the "sleep" movement of bean leaves and the circular circumnutation of both upward-growing cotyledons and downward-growing radicles, comparing the latter to "a burrowing animal such as a mole" seeking to penetrate the ground perpendicularly, moving from side to side to avoid stones. Indeed, beans were as much a research staple as kitchen staple for Darwin.

PINGUICULA Gesneri . L.B.

g. D. Ehret. pinxit.
1767.

Pinguicula vulgaris. Water and bodycolor on vellum by Georg Dionysius Ehret,
in *Flowers, Moths, Butterflies and Shells.*

Pinguicula
BUTTERWORT

—— ...... ——

LENTIBULARIACEAE—BUTTERWORT FAMILY

INSECTIVOROUS PLANTS

Pinguicula grows in low-nutrient moist environments, similar to other carnivorous plants. The rosette-forming light-green leaves glisten with luscious glands, inspiring the Latin name, which refers to fat (*pinguis*), and the common name butterwort for the buttery-looking surface that attracts insects like fly-paper. With approximately ninety species in the genus, several are native to Europe. Darwin was able to observe and work his experiments on three of them, primarily the common circumboreal species *P. vulgaris*.

Butterworts were a late addition to Darwin's insectivorous plant project—he only became aware of their fly-paper qualities in spring of 1874, from family friend William Marshall. He swung into action but knew it would delay publication of his book. "I am now hard at work getting my book on Drosera &c. ready for Printers," he wrote to Asa Gray, at Harvard, "but it will take some time for I am always finding out new points to observe … Day before Yesterday I found out that Pinguicula digests and then absorbs animal matter; I know that this holds good for albumen, gelatin and insects, but I am now in the midst of my observations."[120]

Movement of the leaves of *Pinguicula* had never been suspected or described before Darwin's research, and it enhanced his fascination with the animal-like movement of plants and their digestive abilities. Marshall and his sister Theodora, along with other contacts, friends, and family (including Darwin's future daughter-in-law, Amy Ruck), gladly sent him leaves with insects and seeds trapped upon them, leading Darwin to declare that *Pinguicula* is "not only *insectivorous* but *graminvorous* and *granivorous!*"[121]

Glands on the leaves secrete mucilage with digestive enzymes, and the more the glands are excited, the more mucilage is produced. With live plants under his care, Darwin observed the inward curling of the leaf margins, an action that moves insects onto more glands, triggering more secretion. The curling also pools the digestive enzymes in the concavity, preventing them from leaking off the edges or getting washed away in the rain. Darwin marveled at how he could entice the leaves to respond, timing them as they slowly curled their edges over fragments of proffered flies, roast meat, assorted seeds, pollen, albumen, and other treats. He also measured the roots, their scant size reinforcing his idea that they provide little nutrition for the plant: "We may, therefore, conclude that *Pinguicula vulgaris*, with its small roots, is not only supported to a large extent by the extraordinary number of insects which it habitually captures, but likewise draws some nourishment from the pollen, leaves, and seeds of other plants which often adhere to its leaves. It is therefore partly a vegetable as well as an animal feeder."[122]

[From: *Insectivorous Plants* (2nd ed., 1888)]

Pinguicula vulgaris—This plant grows in moist places, generally on mountains. It bears on an average eight, rather thick, oblong, light green leaves, having scarcely any footstalk. A full-sized leaf is about 1½ inch in length and ¾ inch in breadth. The young central leaves are deeply concave and project upwards; the older ones towards the outside are flat or convex and lie close to the ground, forming a rosette from 3 to 4 inches in diameter. The margins of the leaves are incurved. Their upper surfaces are thickly covered with two sets of glandular hairs, differing in the size of the glands and in the length of their pedicels. The larger glands have a circular outline as seen from above and are of moderate thickness; they are divided by radiating partitions into sixteen cells, containing light-green, homogeneous fluid. They are supported on elongated, unicellular pedicels (containing a nucleus with a nucleolus) which rest on slight prominences. The small glands differ only in being formed of about half the number of cells, containing much paler fluid, and supported on much shorter pedicels. Near the midrib, towards the base of the leaf, the pedicels are multicellular, are longer than elsewhere, and bear smaller glands. All the glands secrete a colourless fluid, which is so viscid that I have seen a fine thread drawn out to a length of 18 inches; but the fluid in this case was secreted by a gland which had been excited. The edge of the leaf is translucent and does not bear any glands; and here the spiral vessels, proceeding from the midrib, terminate in cells marked by a spiral line, somewhat like those within the glands of *Drosera*. ...

A friend sent me on June 23 thirty-nine leaves from North Wales, which were selected owing to objects of some kind adhering to them. Of these leaves, thirty-two had caught 142 insects, or on an average 4.4 per leaf, minute fragments of insects not being included. Besides the insects, small leaves belonging to four different kinds of plants, those of *Erica tetralix* being much the commonest, and three minute seedling plants, blown by the wind, adhered to nineteen of the leaves. One had caught as many as ten leaves of the *Erica*. Seeds or fruits, commonly of *Carex* and one of *Juncus*, besides bits of moss and other rubbish, likewise adhered to six of the thirty-nine leaves. The same friend, on June 27, collected nine plants bearing seventy-four leaves, and all of these, with the exception of three young leaves, had caught insects; thirty insects were counted on one leaf, eighteen on a second, and sixteen on a third. ...

We thus see that numerous insects and other objects are caught by the viscid leaves; but we have no right to infer from this fact that the habit is beneficial to the plant, any more than in the before given case of the *Mirabilis*, or of the horse-chestnut. But it will presently be seen that dead insects and other nitrogenous bodies excite the glands to increased secretion; and that the secretion then becomes acid and has the power of digesting animal substances, such as albumen, fibrin, &c. Moreover, the dissolved nitrogenous matter is absorbed by the glands, as shown by their limpid contents being aggregated into slowly moving granular masses of protoplasm. The same results follow when insects are naturally captured, and as the plant lives in poor soil and has small roots, there can be no doubt that it profits by its power of digesting and absorbing matter from the prey which it habitually captures in such large numbers. It will, however, be convenient first to describe the movements of the leaves.

That such thick, large leaves as those of *Pinguicula vulgaris* should have the power of curving inwards when excited has never even been suspected. It is necessary to select for experiment leaves with their glands secreting freely, and which have been prevented from capturing many insects; as old leaves, at least those growing in a state of nature, have their margins already curled so much inwards that they exhibit little power of movement, or move very slowly. ...

We learn from [experiment] that the margins of the leaves curl inwards when excited by the mere pressure of objects not yielding any soluble matter, by objects yielding such matter, and by some fluids-namely an infusion of raw meat and a week solution of carbonate of ammonia. A stronger solution of two grains of this salt to an ounce of water, though exciting copious secretion, paralyses the leaf. Drops of water and of a solution of sugar or gum did not cause any movement. Scratching the surface of the leaf for some minutes produced no effect. Therefore, as far as we at present know, only two causes—namely slight continued pressure and the absorption of nitrogenous matter—excite movement. It is only the margins of the leaf which bend, for the apex never curves towards the base. The pedicels of the glandular hairs have no power of movement. ...

... We have seen that when large bits of meat, or of sponge soaked in the juice of meat, were placed on a leaf, the margin was not able to embrace them, but, as it became incurved, pushed them very slowly towards the middle of the leaf, to a distance from the outside of fully .1 of an inch (2.54 mm.), that is, across between one-third and one-fourth of the space between the edge and midrib. Any object, such as a moderately

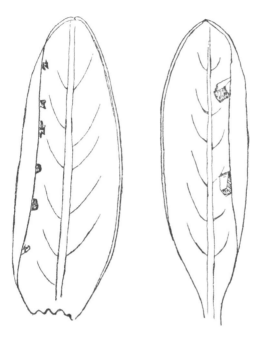

Leaf margin inflected over a row of small flies, left,
and against two bits of meat, right

sized insect, would thus be brought slowly into contact with a far larger number of glands, inducing much more secretion and absorption, than would otherwise have been the case. That this would be highly service-able to the plant, we may infer from the fact that *Drosera* has acquired highly developed powers of movement, merely for the sake of bringing all its glands into contact with captured insects. So again, after a leaf of *Dionaea* has caught an insect, the slow pressing together of the two lobes serves merely to bring the glands on both sides into contact with it, causing also the secretion charged with animal matter to spread by capillary attrac-tion over the whole surface. In the case of *Pinguicula*, as soon as an insect has been pushed for some little distance towards the midrib, immediate re-expansion would be beneficial, as the margins could not capture fresh prey until they were unfolded.

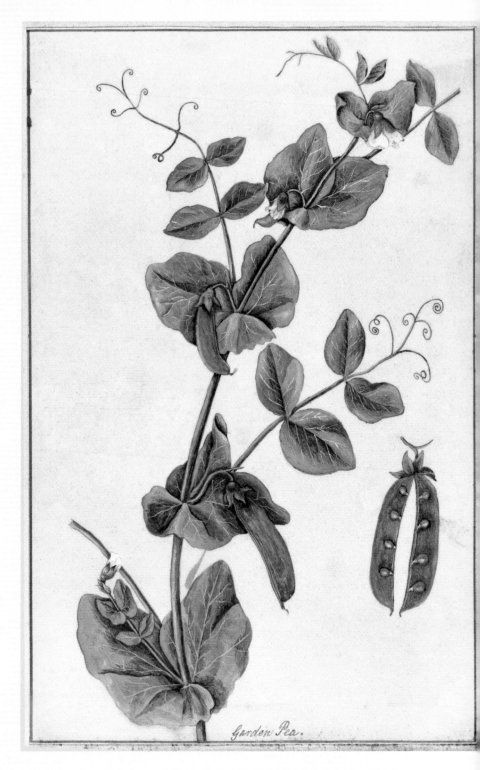

Pisum sativum. Watercolor by Elizabeth Blackwell, *A Curious Herbal.*

Pisum

GARDEN PEA

———— ...... ————

FABACEAE—PEA FAMILY

VARIATION, CROSS AND SELF-FERTILIZATION

One indication that a plant has been associated with people for a long time is that its genus is the name for that plant in some ancient language—so it is with *pisum*, the Latin word for the garden pea. Pea plants, grown for their exceptionally nutritious pods and seeds, are herbaceous annuals with compound leaves ending in tendrils that make them good climbers. This staple food source, in particular *Pisum sativum*, is thought to have originated in the circum-Mediterranean region and Middle East, where dried peas have been consumed for some 7000 years. Of the many varieties available in Victorian England, Darwin grew forty-one in his garden at Down—to observe and experiment with, and, of course, to eat; Emma Darwin's recipe book includes a hearty "pea soop."[123]

Her husband's studies with peas covered many subjects, published in several books and articles over the years, beginning with his earliest investigations into the question of transmutation, or species change, in the late 1830s. With the help of his father's gardener, John Abberley, Darwin featured *Pisum* in some of his very first experiments, which tested whether certain long-established plant varieties would breed true and resist crossing with related varieties. This would mean such varieties bore the hallmark of separate species, consistent with Darwin's growing conviction that varieties were but incipient species.*[124] In one test, Abberley planted rows of different pea varieties then planted out the resulting seeds of each. In another,

Darwin asked him to plant individual pea plants mixed in with beans. The results in both cases supported the concept that the pea varieties bred true.**[125]

$$\left[\begin{array}{c} \text{From: } \textit{The Variation of Animals and Plants Under} \\ \textit{Domestication, Volume 1 (2nd ed., 1875)} \end{array} \right]$$

Pea (Pisum sativum).— ... The varieties of the common garden-pea are numerous and differ considerably from one another. For comparison I planted at the same time forty-one, English and French varieties. They differed greatly in height,—namely from between 6 and 12 inches to 8 feet, —in manner of growth, and in period of maturity. Some differ in general aspect even while only two or three inches in height. The stems of the *Prussian* pea are much branched. The tall kinds have larger leaves than the dwarf kinds, but not in strict proportion to their height:—*Hair's Dwarf Monmouth* has very large leaves, and the *Pois nain hatif,* and the moderately tall *Blue Prussian*, have leaves about two-thirds of the size of the tallest kind. In the *Danecroft* the leaflets are rather small and a little pointed; in the *Queen of Dwarfs* rather rounded; and in the *Queen of England* broad and large. In these three peas the slight differences in the shape of the leaves are accompanied by slight differences in colour, in the *Pois géant sans parchemin*, which bears purple flowers, the leaflets in the young plant are edged with red; and in all the peas with purple flowers the stipules are marked with red.

---

* (previous) Darwin's interest in this question was piqued when reading a paper by the renowned plant breeder Rev. William Herbert, prompting Darwin to write Herbert with a list of plant breeding questions and undertake experiments of his own.

** An entry in Darwin's "Questions & Experiments" notebook reads, "Get Abberley to plant single Peas, Kidney Bean and Bean, intertwined ... in reference to what Mr Herbert observe[ed] on this subject."

In the different varieties, one, two, or several flowers in a small cluster, are borne on the same peduncle; and this is a difference which is considered of specific value in some of the Leguminosae. In all the varieties, the flowers closely resemble each other except in colour and size. They are generally white, sometimes purple, but the colour is inconstant even in the same variety. ...

The pods and seeds, which with natural species afford such constant characters, differ greatly in the cultivated varieties of the pea; and these are the valuable, and consequently the selected parts. *Sugar peas*, or *Pois sans parchemin*, are remarkable from their thin pods, which, whilst young, are cooked and eaten whole; and in this group ... it is the pod which differs most; thus *Lewis's Negro-podded pea* has a straight, broad, smooth, and dark-purple pod, with the husk not so thin as in the other kinds; the pod of another variety is extremely bowed; that of the *Pois géant* is much pointed at the extremity; and in the variety "*à grands cosses*" the peas are seen through the husk in so conspicuous a manner that the pod, especially when dry, can hardly at first be recognised as that of a pea.

In the ordinary varieties, the pods also differ much in size;—in colour, that of *Woodford's Green Marrow* being bright-green when dry, instead of pale brown, and that of the purple-podded pea being expressed by its name;—in smoothness, that of *Danecroft* being remarkably glossy, whereas that of the *Ne plus ultra* is rugged; in being either nearly cylindrical, or broad and flat;—in being pointed at the end, as in *Thurston's Reliance*, or much truncated, as in the *American Dwarf*. In the *Auvergne pea*, the whole end of the pod is bowed upwards. In the *Queen of Dwarfs* and in *Scimitar peas*, the pod is almost elliptic in shape. I here give drawings of the four most distinct pods produced by the plants cultivated by me.

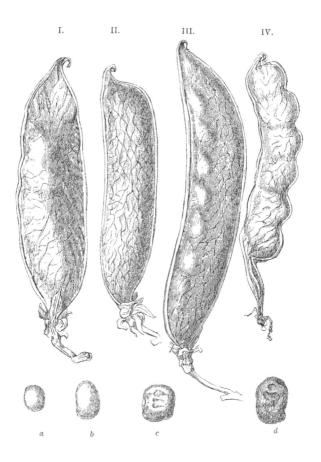

Pods and Peas. I. Queen of Dwarfs. II. American Dwarf. III. Thurston's Reliance. IV. Pois Géant sans parchemin. *a*. Dan O'Rourke Pea. *b*. Queen of Dwarfs Pea. *c*. Knight's Tall White Marrow. *d*. Lewis's Negro Pea.

With respect to the varieties not naturally intercrossing, I have ascertained that the pea, which in this respect differs from some other Leguminosae, is perfectly fertile without the aid of insects. Yet I have seen humble-bees whilst sucking the nectar depress the keel-petals, and become so thickly dusted with pollen, that it could hardly fail to be left on the stigma of the next flower which was visited. Nevertheless, distinct varieties growing closely together rarely cross; and I have reason to believe that this is due to their stigmas being prematurely fertilised in this country by pollen from the same flower. The horticulturists who raise seed-peas are thus enabled to plant distinct varieties close together without any bad consequences; and it is certain, as I have myself found, that true seed may be saved during at least several generations under these circumstances.

[From: *The Effects of Cross and Self Fertilisation in the Vegetable Kingdom* (2nd ed., 1878)]

The common pea is perfectly fertile when its flowers are protected from the visits of insects; I ascertained this with two or three different varieties, as did Dr. Ogle with another. But the flowers are likewise adapted for cross-fertilisation; Mr. Farrer specifies the following points, namely: "The open blossom displaying itself in the most attractive and convenient position for insects; the conspicuous vexillum; the wings forming an alighting place; the attachment of the wings to the keel, by which any body pressing on the former must press down the latter; the staminal tube enclosing nectar, and affording by means of its partially free stamen with apertures on each side of its base an open passage to an insect seeking the nectar; the moist and sticky pollen placed just where it will be swept out of the apex of the keel against the entering insect; the stiff elastic style so placed that on a pressure being applied to the keel it will be pushed upwards out of the keel; the hairs on the style placed on that side of the style only on which there is space for the pollen, and in such a direction as to sweep it out; and the stigma so placed as to meet an entering insect,—all these become correlated parts of one elaborate mechanism, if we suppose that the fertilisation of these flowers is effected by the carriage of pollen from one to the other." Notwithstanding these manifest provisions for cross-fertilisation, varieties which have been cultivated for very many successive generations in close proximity, although flowering at the same time, remain pure. ...

It is remarkable, considering that the flowers secrete much nectar and afford much pollen, how seldom they are visited by insects either in England or, as H. Müller remarks, in North Germany. I have observed the flowers for the last thirty years, and in all this time have only thrice seen bees of the proper kind at work (one of them being *Bombus muscorum*), such as were sufficiently powerful to depress the keel, so as to get the undersides of their bodies dusted with pollen. These bees visited several flowers and could hardly have failed to cross-fertilise them. Hive-bees and other small kinds sometimes collect pollen from old and already fertilised flowers, but this is of no account. The rarity of the visits of efficient bees to this exotic plant is, I believe, the chief cause of the varieties so seldom inter-crossing. That a cross does occasionally take place, as might be expected from what has just been stated, is certain, from the recorded cases of the direct action of the pollen of one variety on the seed-coats of another. ...

We understand today that Abberley's results stem from the fact that pea flowers typically self-pollinate. Varieties will readily cross when hand-pollinated, as Austrian monk Gregor Mendel showed. Like Darwin, Mendel chose peas for his experiments because he could grow them easily and readily observe traits of interest. He cross-bred tall and dwarf pea varieties, green and yellow peas, purple and white flowers, wrinkled and smooth peas, and observed the resulting offspring—experiments that led to the foundation of modern genetics. Contrary to popular belief, Darwin was unaware of the monk's pea-crossing papers, and even if he *had* read them he probably would not have appreciated their significance—that insight came forty years later, when biologists could understand Mendel in light of late nineteenth-century discoveries of chromosomes and meiosis.

Great Cowslip.

Primula veris. Water and bodycolor on vellum by Dame Ann Hamilton,
Drawings of Plants.

Primula

PRIMROSE

———— ...... ————

PRIMULACEAE—PRIMROSE FAMILY

FORMS OF FLOWERS, POLLINATION

Primula is a large genus, with nearly 500 species and countless cultivars, all low-growing perennial herbs, with tufted rosettes of leaves highlighted by showy flowers arranged as umbels on naked stalks. Darwin worked on several species including three common in his woods and meadows, *P. veris* (cowslip), *P. vulgaris* (primrose), and the uncommon *P. elatior* (oxlip), and five other species including *P. sinensis* from China. One significant insight from his experiments involved oxlips, long regarded as the hybrid offspring of primroses and cowslips. His early ideas about the plants were published in the first edition of *Origin*,[126] but after experiments crossbreeding them, he determined that they were true species rather than varieties, earning him further respect from botanists. The plants were omitted from later editions of *Origin* when he realized his mistake.

[**From: *On the Origin of Species* (1st ed., 1859)**]

Many of the cases of strongly-marked varieties or doubtful species well deserve consideration; for several interesting lines of argument, from geographical distribution, analogical variation, hybridism, &c., have been brought to bear on the attempt to determine their rank. I will here give only a single instance,—the well-known one of the primrose and cowslip, or *Primula veris* and elatior. These plants differ considerably in appearance;

they have a different flavour and emit a different odour; they flower at slightly different periods; they grow in somewhat different stations; they ascend mountains to different heights; they have different geographical ranges; and lastly, according to very numerous experiments made during several years by that most careful observer Gärtner, they can be crossed only with much difficulty. We could hardly wish for better evidence of the two forms being specifically distinct. On the other hand, they are united by many intermediate links, and it is very doubtful whether these links are hybrids; and there is, as it seems to me, an overwhelming amount of experimental evidence, showing that they descend from common parents and consequently must be ranked as varieties.

By 1860, Darwin had become curious about the two distinct kinds of flowers found in primroses, one with long pistils and short stamens (so-called "pin" flowers, which Darwin called "female") and the other just the opposite, with short pistils and long stamens ("thrum," or "male" flowers to Darwin). He knew that botanists and other observers were aware of this curious dimorphism, termed heterostyly today, but also that no one considered their purpose. At first Darwin thought the morphs were evolving into separate sexes, that he fortuitously caught them in the very process of diverging. But in 1860, soon after the publication of *Origin*, he began crossing experiments, beginning with putting his kids to work collecting the different morphs—281 "male" flowers and 241 "female" in one productive haul. He observed and measured pollen grains of each morph and did a series of tests crossing pollen between them, establishing "legitimate marriages" that yielded seed, where the stigma of one form is pollinated by pollen from stamens of another form, and fruitless "illegitimate marriages" when pollen from the same plant was transferred its own stigma. He grew plants for several generations this way and counted the seeds produced by each morph, hypothesizing that the "male" flowers would produce fewer seeds since they were becoming less and less "female."

He found just the opposite, however—"male" flowers produced *more* seeds, not fewer—thus running headlong into what his friend Thomas Henry Huxley called the "great tragedy of science," namely, "the slaying of a beautiful hypothesis by an ugly fact."[127]

Darwin ultimately abandoned the idea that the flower morphs were evolving separate sexes and came to see them as an adaptation to promote outcrossing between individuals as much as possible. The results of his observations and experiments were first presented at the Linnean Society of London in 1862, and then in *Forms of Flowers* in 1877, with more experiments described in *Cross and Self Fertilisation*. Darwin's research on primulas began his great interest in other genera and families with heterostyly; so much so that he wrote in his autobiography, "No little discovery of mine ever gave me so much pleasure as the making out the meaning of heterostyled flowers."[128]

> **From: "On the two forms, or dimorphic condition, in the species of *Primula*, and on their remarkable sexual relations."** *Journal of the Proceedings of the Linnean Society of London* (Botany) 6: 77–96. (1862)

If a large number of Primroses or Cowslips (*P. vulgaris* and *veris*) be gathered, they will be found to consist, in about equal numbers, of two forms, obviously differing in the length of their pistils and stamens. Florists who cultivate the Polyanthus and Auricula are well aware of this difference, and call those which display the globular stigma at the mouth of the corolla "pin-headed" or "pin-eyed," and those which display the stamens "thrum-eyed." I will designate the two forms as long-styled and short-styled. Those botanists with whom I have spoken on the subject have looked at the case as one of mere variability, which is far from the truth.

In the Cowslip, in the long-styled form, the stigma projects just above the tube of the corolla and is externally visible; it stands high above the anthers, which are situated halfway down the tube, and cannot be easily seen. In the short-styled form, the anthers are attached at the mouth of the tube, and therefore stand high above the stigma; for the pistil is short, not rising above halfway up the tubular corolla. The corolla itself is of a different shape in the two forms, the throat or expanded portion above the attachment of the anthers being much longer in the long-styled than in the short-styled form. Village children notice this difference, as they can best make necklaces by threading and slipping the corollas of the long-styled flowers into each other. But there are much more important differences. The stigma in the long-styled plants is globular, in the short-styled it is depressed on the summit, so that the longitudinal axis of the former is sometimes nearly double that of the latter. The shape, however, is in some degree variable; but one difference is persistent, namely, that the stigma of the long-styled is much rougher: in some specimens carefully compared, the papillae which render the stigmas rough were in the long-styled form from twice to thrice as long as in the short-styled. There is another and more remarkable difference, namely, in the size of the pollen-grains. I measured with the micrometer many specimens, dry and wet, taken from plants growing in different situations, and always found a palpable difference. ...

There is also a difference in shape, the grains from the short-styled plants being nearly spherical, those from the long-styled being oblong with the angles rounded; this difference in shape disappears when the grains are distended with water. Lastly, as shall presently see, the short-styled plants produce more seed than the long-styled.

To sum up the differences:—The long-styled plants have a much longer pistil, with a globular and much rougher stigma, standing high above the anthers. The stamens are short; the grains of pollen smaller and oblong in shape. The upper half of the tube of the corolla is more expanded. The number of seeds produced is smaller.

The short-styled plants have a short pistil, half the length of the tube of the corolla, with a smooth depressed stigma standing beneath the anthers. The stamens are long; the grains of pollen are spherical and larger. The tube of the corolla is of the same diameter till close to its upper end. The number of seeds produced is larger.

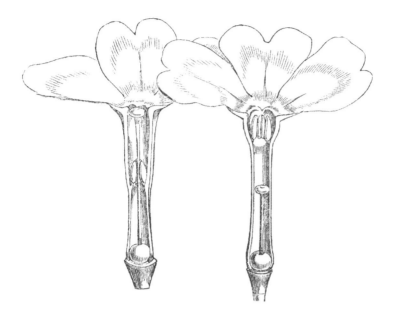

Long-styled, left. Short-styled, right.

I have examined a large number of flowers; and though the shape of the stigma and the length of the pistil vary, especially in the short-styled form, I have never seen any transitional grades between the two forms. There is never the slightest doubt under which form to class a plant. I have never seen the two forms on the same plant. I marked many Cowslips and Primroses, and found, the following year, that all retained the same character, as did some in my garden which flowered out of their proper season in the autumn. ... An excellent proof of the permanence of the two forms is seen in nursery gardens, where choice varieties of the Polyanthus are propagated by division; and I found whole beds of several varieties, each consisting exclusively of the one or the other form. The two forms exist in the wild state in about equal numbers: I collected from several different stations, taking every plant which grew on each spot, 522 umbels; 241 were long-styled, and 281 short-styled. No difference in tint or size could be perceived in the two great masses of flowers.

I examined many cultivated Cowslips (*P. veris*) or Polyanthuses, and Oxlips; and the two forms always presented the same differences, including the same relative difference in the size of the pollen-grains. ...

The first idea which naturally occurred was, that the species were tending towards a dioicous condition; that the long-styled plants, with their rougher stigmas, were more feminine in nature, and would produce more seed; that the short-styled plants, with their long stamens and larger pollen-grains, were more masculine in nature. Accordingly, in 1860, I marked some Cowslips of both forms growing in my garden, and others growing in an open field, and others in a shady wood, and gathered and weighed the seed. In each of these little lots the short-styled plants yielded, contrary to my expectation, most seed. ...

In 1861, I tried the result in a fuller and fairer manner. I transplanted in the previous autumn a number of wild plants into a large bed in my garden, treating them all alike. ...

The season was much better this year than the last, and the plants grew in good soil, instead of in a shady wood or struggling with other plants in the open field; consequently the actual produce of seed was considerably greater. Nevertheless, we have the same relative result; for the short-styled plants produced more seed than the long-styled in the proportion of three to two. ...

For the sake of clearness, the general result is given in the following diagram, in which the dotted lines with arrows represent how in the four unions pollen has been applied.

We here have a case new, as far as I know, in the animal and vegetable kingdoms. We see the species of *Primula* divided into two sets or bodies, which cannot be called distinct sexes, for both are hermaphrodites; yet they are to a certain extent sexually distinct, for they require for perfect fertility reciprocal union. They might perhaps be called sub-dioicous hermaphrodites. As quadrupeds are divided into two nearly equal bodies of different sexes, so here we have two bodies, approximately equal in number, differing in their sexual powers and related to each other like males

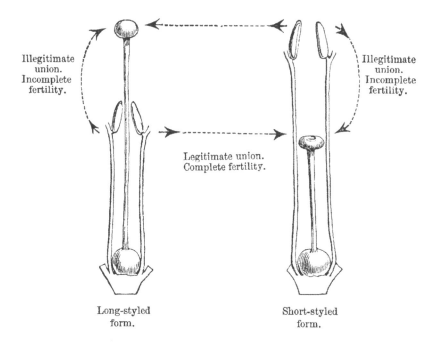

Illegitimate union.
Incomplete fertility.

Illegitimate union.
Incomplete fertility.

Legitimate union.
Complete fertility.

Long-styled form.

Short-styled form.

and females. There are many hermaphrodite animals which cannot fertil-ize themselves, but must unite with another hermaphrodite: so it is with numerous plants; for the pollen is often mature and shed, or is mechani-cally protruded, before the flower's own stigma is ready; so that these her-maphrodite flowers absolutely require for their sexual union the presence of another hermaphrodite. But in *Primula* there is this wide difference, that one individual Cowslip, for instance, though it can with mechanical aid imperfectly fertilize itself, for full fertility must unite with another indi-vidual; but it cannot unite with any individual in the same manner as an hermaphrodite Snail or Earth-worm can unite with any other one Snail or Earth-worm; but one form of the Cowslip, to be perfectly fertile, must unite with one of the other form, just as a male quadruped must and can unite only with a female.

Pulmonaria officinalis. Watercolor by Elizabeth Wharton, *British Flowers*.

┌─────────────────────────────────────┐

Pulmonaria

LUNGWORT

──── ⋯⋯⋯ ────

BORAGINACEAE—BORAGE FAMILY

└─────────────────────────────────────┘

FORMS OF FLOWERS, POLLINATION

The Latin name as well as the common name of *Pulmonaria* refers to the lung, in a nod to herbalists of old who believed that the spotted leaves of *Pulmonaria officinalis* cure diseased and ulcerated lungs. This supposed relationship between plants and people is based on the medieval "Doctrine of Signatures," which held that plants or plant parts resembling any part of human anatomy possessed curative properties for that body part.

There are around fifteen species of *Pulmonaria*, native to Europe and beyond, with many cultivars grown in gardens throughout the temperate world. They are perennial herbs with basal leaf rosettes and attractive flower racemes in a spiraled or "scorpioid" arrangement, often initially blue in color and transitioning to pink after pollination. Along with *Primula*, *Linum*, and other plants that Darwin studied, they present a nice example of flower dimorphism, bearing heterostylous flowers with different lengths of stamens and pistils that function to facilitate cross pollination.

Darwin's eldest son William discovered floral dimorphism in a population of narrow-leaved lungwort (*Pulmonaria angustifolia*) on the Isle of Wight in spring of 1863. His father was delighted, eager to expand his studies of heterostyly and inter-fertility in primroses and flax.[129] Well-stocked by William with seedlings of *P. angustifolia* and seeds of another species, the common blue lungwort (*P. officinalis*), Darwin experimented with both plants the following year. When he began his work he noted that *P. angustifolia* and *P. officinalis* are so similar that some botanists considered them as "mere varieties" of the same species. Through cross-pollination tests, however, he determined that they were truly distinct species. He expected that the flower morphs would be cross-fertile and self-sterile, consistent with other hetero-stylous plants he had studied. But while the long-styled form of *P. angustifolia* was self-sterile, he was surprised to find the short-styled form to be self-fertile to a remarkable degree.

He commented on this to Asa Gray in the United States, saying, "Did I ever tell you that a year or two ago I ascertained that *Pulmonaria* offers a curious case. The long-styled form being absolutely sterile with its own pollen, whilst the short-styled is almost perfectly fertile with its own pollen." This puzzling departure from the more typical result when experimentally self-pollinating within morphs of heterostylous plants—which tend to yield sterile and often stunted offspring—prompted Darwin to declare, "This seems to me a very curious fact."[130]

[**From: *The Different Forms of Flowers on Plants of the Same Species* (1877)**]

Pulmonaria angustifolia.—Seedlings of this plant, raised from plants growing wild in the Isle of Wight, were named for me by Dr. Hooker. It is so closely allied to the last species, differing chiefly in the shape and spotting of the leaves, that the two have been considered by several emi-nent botanists—for instance, Bentham—as mere varieties. But, as we shall presently see, good evidence can be assigned for ranking them as distinct. Owing to the doubts on this head, I tried whether the two would mutually fertilise one another. Twelve short-styled flowers of *P. angustifolia* were legitimately fertilised with pollen from long-styled plants of *P. officinalis* (which, as we have just seen, are moderately self-fertile), but they did not produce a single fruit. Thirty-six long-styled flowers of *P. angustifolia* were also illegitimately fertilised during two seasons with pollen from the long-styled *P. officinalis*, but all these flowers dropped off unimpregnated. Had the plants been mere varieties of the same species these illegitimate crosses would probably have yielded some seeds, judging from my success in illegitimately fertilising the long-styled flowers of *P. officinalis*; and the twelve legitimate crosses, instead of yielding no fruit, would almost certainly have yielded a considerable number, namely, about nine. ... Therefore *P. officinalis* and *angustifolia* appear to be good and distinct species, in conformity with other important functional differences between them, immediately to be described.

The long-styled and short-styled flowers of *P. angustifolia* differ from one another in structure in nearly the same manner as those of *P. officinalis*. But in the accompanying figure a slight bulging of the corolla in the long-styled form, where the anthers are seated, has been overlooked. My son William, who examined a large number of wild plants in the Isle of Wight, observed that the corolla, though variable in size, was generally larger in the long-styled flowers than in the short-styled; and certainly the largest corollas of all were found on the long-styled plants, and the smallest on the short-styled. Exactly the reverse occurs, according to Hildebrand, with *P. officinalis*. Both the pistils and stamens of *P. angustifolia* vary much in length ... so that the stigma in the one form does not stand on a level with the anthers in the other. The long-styled pistil is sometimes thrice as long as that of the short-styled; but from an average of ten measurements of both, its length to that of the short-styled was as 100 to 56. The stigma varies in being more or less, though slightly, lobed.

My son collected in the Isle of Wight on two occasions 202 plants, of which 125 were long-styled and 77 short-styled; so that the former were the more numerous. On the other hand, out of 18 plants raised by me from seed, only 4 were long-styled and 14 short-styled. The short-styled plants seemed to my son to produce a greater number of flowers than the long-styled; and he came to this conclusion before a similar statement had been published by Hildebrand with respect to *P. officinalis*. My son gathered ten branches from ten different plants of

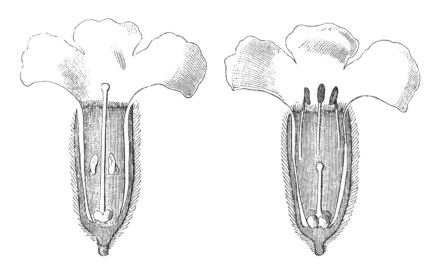

Pulmonaria angustifolia. Long-styled, left.
Short-styled, right.

both forms and found the number of flowers of the two forms to be as 100
to 89, 190 being short-styled and 169 long-styled. With *P. officinalis* the
difference, according to Hildebrand, is even greater, namely, as 100 flowers
for the short-styled to 77 for the long-styled plants. ...

The fertility of the two legitimate unions to that of the two illegitimate
together is as 100 to 35, judged by the proportion of flowers which produced fruit; and as 100 to 32, judged by the average number of seeds per
fruit. But the small number of fruit yielded by the 18 long-styled flowers
in the first line was probably accidental, and if so, the difference in the
proportion of legitimately and illegitimately fertilised flowers which yield
fruit is really greater than that represented by the ratio of 100 to 35. The
18 long-styled flowers illegitimately fertilised yielded no seeds,—not even
a vestige of one. Two long-styled plants which were placed under a net
produced 138 flowers, besides those which were artificially fertilised, and
none of these set any fruit; nor did some plants of the same form which

were protected during the next summer. Two other long-styled plants were left uncovered (all the short-styled plants having been previously covered up), and humble-bees, which had their foreheads white with pollen, incessantly visited the flowers, so that their stigmas must have received an abundance of pollen, yet these flowers did not produce a single fruit. We may therefore conclude that the long-styled plants are absolutely barren with their own-form pollen, though brought from a distinct plant. In this respect they differ greatly from the long-styled English plants of *P. officinalis* which were found by me to be moderately self-fertile; but they agree in their behavior with the German plants of *P. officinalis* experimented on by Hildebrand. ...

The great difference in the fertility of the long and short-styled flowers, when both are illegitimately fertilised, is a unique case, as far as I have observed with heterostyled plants. The long-styled flowers when thus fertilised are utterly barren, whilst about half of the short-styled ones produce capsules, and these include a little above two-thirds of the number of seeds yielded by them when legitimately fertilised. The sterility of the illegitimately fertilised long-styled flowers is probably increased by the deteriorated condition of their pollen; nevertheless, this pollen was highly efficient when applied to the stigmas of the short-styled flowers.

Salvia coccinea. Hand-colored engraving by William Jackson Hooker,
from *The Botanical Magazine* 1828 55: 2864.

Salvia

SAGE

——— ...... ———

LAMIACEAE—MINT FAMILY

FORMS OF FLOWERS, POLLINATION, CROSS AND SELF-FERTILIZATION

With more than 700 species and innumerable cultivars, salvias are prized for their bold colorful flowers and fragrant foliage. Herbalists of old found them attractive for another reason—the Latin name is derived from *salvare*, "to heal," and *salvus*, "uninjured," in reference to the medicinal uses of the foliage in compresses for wounds and other treatments dating back to antiquity. Darwin, however, was interested in the *plant's* wounds—he noted that bumblebees often avoid entering the mouth of *Salvia* flowers, preferring to bore into the flower base close to the spot where the nectar lies hidden within the corolla— nectar robbery, not very helpful for pollination.

[**From: "Humble-bees."** *Gardeners' Chronicle* (21 August 1841, p. 550)]

Perhaps some of your readers may like to hear a few more particulars about the humble-bees which bore holes in flowers, and thus extract the nectar. This operation has been performed on a large scale in the Zoological Gardens:—Near the refection-house there is a fine bed of *Stachys coccinea*, every flower in which has one, and sometimes two, small irregular slits, or orifices, on the upper side of the corolla near its base. I observed some plants of Marvel of Peru, and of *Salvia coccinea*, with holes in similar positions; but in *Salvia Grahami* they were without exception cut through the calyx.

Those bees that do enter the flower in the way the plants preferred them to, so to speak, encounter a wonderful pollination adaptation: two stamens fused such that one protrudes from the other like a handle, acting as a lever when the insect pushes past it to get to the nectary, tipping the attached pollen-laden stamen down onto its back where pollen is smeared. It was just the sort of adaptation that thrilled Darwin, who declared the fused stamens "as perfect a structure as can be found in any orchid."[131]

This mechanism was first described by German botanist Friedrich Hildebrand, who wrote Darwin in 1864 that his research into the pollination mechanisms of *Salvia*, *Pulmonaria*, *Linum*, and other plants was inspired by Darwin's orchid book.[132] (Hildebrand, whom Darwin cites extensively in his 1876 book *Cross and Self Fertilisation*, coined the term "heterostyly," which Darwin thought was an improvement over his own terms "di- and trimorphic" to describe these flowers.) Bees in a hurry didn't seem to appreciate the biomechanical beauty of *Salvia*, however perfect Darwin found it. He observed that flowers of the vivid scarlet sage (*Salvia coccinea*) were often perforated with one or two slits on the upper side of the corolla near the base, and both the calyx and corolla of blackcurrant or Graham's sage, now *Salvia microphylla*, were invariably perforated by what he labeled "pick-pocket bees."

Wondering to what extent seeds were produced when insects failed to pollinate the flowers, he worked under a net, artificially self-pollinating some and leaving others alone. It turns out that *Salvia coccinea* does not need the bees to set seed, though they do better when pollinated as described in Darwin's study. How did this compare with seed-set in outcrossed flowers? To Darwin's surprise, plenty of seeds were produced regardless of how they were pollinated, but then, after growing them out, he found that the cross-fertilized plants grew significantly larger and produced nearly twice as many flowers as the self-pollinated ones—yet another demonstration of the benefits of outcrossing.

From: *The Effects of Cross and Self Fertilisation in the Vegetable Kingdom* (2nd ed., 1878)

Salvia coccinea.—This species, unlike most of the others in the same genus, yields a good many seeds when insects are excluded. I gathered ninety-eight capsules produced by flowers spontaneously self-fertilised under a net, and they contained on an average 1.45 seeds, whilst flowers artificially fertilised with their own pollen, in which case the stigma will have received plenty of pollen, yielded on an average 3.3 seeds, or more than twice as many. Twenty flowers were crossed with pollen from a distinct plant, and twenty-six were self-fertilised. There was no great difference in the proportional number of flowers which produced capsules by these two processes, or in the number of the contained seeds or in the weight of an equal number of seeds.

Seeds of both kinds were sown rather thickly on opposite sides of three pots. When the seedlings were about 3 inches in height, the crossed showed a slight advantage over the self-fertilised. When two-thirds grown, the two tallest plants on each side of each pot were measured; the crossed averaged 16.37 inches, and the self-fertilised 11.75 in height; or as 100 to 71. When the plants were fully grown and had done flowering, the two tallest plants on each side were again measured …

… Each of the six tallest crossed plants exceeds in height its self-fertilised opponent; the former averaged 27.85 inches, whilst the six tallest self-fertilised plants averaged 21.16 inches; or as 100 to 76. In all three pots the first plant which flowered was a crossed one. All the crossed plants together produced 409 flowers, whilst all the self-fertilised together produced only 232 flowers; or as 100 to 57. So that the crossed plants in this respect were far more productive than the self-fertilised.

Potatas
or Potadoes.

Solanum tuberosum. Water and bodycolor on vellum by English School artist,
Album of Garden Flowers.

Solanum
NIGHTSHADES

———— ...... ————

SOLANACEAE—NIGHTSHADE FAMILY

VARIATION, CLIMBING PLANTS

The genus *Solanum* has a worldwide distribution, with the greatest
species diversity found in tropical South America. Linnaeus named
the type species *S. nigrum* in 1753, and today more than 1400 species
are recognized, some growing as trees and shrubs while others are
herbs or vines. *Solanum* and other members of the family are of great
economic importance, a group that includes such crops as potatoes,
tomatoes, tobacco, and eggplants, as well as many cultivated ornamentals and medicinal plants.

In his early evolutionary speculations, domestication gave Darwin
insight into the formation of varieties and species, and *Solanum* was
a prime example. In January 1835, traveling aboard the HMS *Beagle*
in the Chonos Archipelago off the Chilean coast, he came upon an
unpromising species of wild potato growing abundantly on Guayteca
Island. "The tallest plant was four feet in height," he later wrote in
his *Journal of Researches*. "The tubers were generally small, but I found
one of an oval shape, two inches in diameter: they resembled in every
respect, and had the same smell as English potatoes; but when boiled
they shrunk much and were watery and insipid."[133] Darwin's potato is
now known as *Solanum ochoanum*, a name honoring the distinguished
Peruvian potato expert Carlos Ochoa, who rediscovered the plant
years later, growing right where Darwin found it.[134] Darwin recalled
this wild species when describing the remarkable diversity of domes-
tic potato varieties in *Variation*.[135]

He went on to grow many varieties in his garden—a double benefit as subjects for scientific study and food on the table. (Potato Rissoles were a favorite.[136]) Comparing the varieties, Darwin noted the traits that varied and those that did not. The stems, leaves, and for the most part the flowers were all far more similar among varieties than were the tubers—unsurprisingly, Darwin argued, since it's the tubers that are under selection by people, not the leaves or flowers.

From: *The Variation of Plants and Animals Under Domestication* (2nd ed., 1875)

Potato (*Solanum tuberosum*).—There is little doubt about the parentage of this plant; for the cultivated varieties differ extremely little in general appearance from the wild species, which can be recognised in its native land at the first glance. The varieties cultivated in Britain are numerous; thus Lawson gives a description of 175 kinds. I planted eighteen kinds in adjoining rows; their stems and leaves differed but little, and in several cases there was as great a difference between the individuals of the same variety as between the different varieties. The flower varied in size, and in colour between white and purple, but in no other respect, except that in one kind the sepals were somewhat elongated. One strange variety has been described which always produces two sorts of flowers, the first double and sterile, the second single and fertile. The fruit or berries also differ, but only in a slight degree. The varieties are liable in very different degree to the attack of the Colorado potato-beetle.

The tubers, on the other hand, present a wonderful amount of diversity. This fact accords with the principle that the valuable and selected parts of all cultivated productions present the greatest amount of modification. They differ much in size and shape, being globular, oval, flattened, kidney-like, or cylindrical. One variety from Peru is described as being quite straight, and at least six inches in length, though no thicker than a man's

finger. The eyes or buds differ in form, position, and colour. The manner in which the tubers are arranged on the so-called roots or rhizomes
is different; thus, in the *gurken-kartoffeln* they form a pyramid with the
apex downwards, and in another variety they bury themselves deep in the
ground. The roots themselves run either near the surface or deep in the
ground. The tubers also differ in smoothness and colour, being externally
white, red, purple, or almost black, and internally white, yellow, or almost
black. They differ in flavour and quality, being either waxy or mealy; in their
period of maturity, and in their capacity for long preservation.

As with many other plants which have been long propagated by bulbs,
tubers, cuttings, &c., by which means the same individual is exposed during
a length of time to diversified conditions, seedling potatoes generally display
innumerable slight differences. Several varieties, even when propagated by
tubers, are far from constant ... Dr. Anderson procured seed from an Irish
purple potato, which grew far from any other kind, so that it could not, at
least in this generation, have been crossed, yet the many seedlings varied in
almost every possible respect, so that "scarcely two plants were exactly alike."
Some of the plants which closely resembled each other above ground produced extremely dissimilar tubers; and some tubers which externally could
hardly be distinguished, differed widely in quality when cooked.

Darwin ran several other garden and greenhouse studies of temperate
and tropical *Solanum* species, ranging from pollination and grafting to
cotyledon movement and climbing habits. In the latter case, he was
especially struck by *Solanum jasminoides* (now *S. laxum*), a cultivated
species from South America with a strong leaf-climbing tendency.
The petioles of this scrambling climber act a bit like tendrils, wrap-
ping themselves about supports, becoming thicker and more robust
in the process. Darwin studied cross-sections of these hefty petioles,
marveling that they can become thicker than even the main stem
itself—a plant "singular" in morphology and physiology.

Solanum jasminoides.— Some of the species in this large genus are twiners; but the present species is a true leaf-climber. A long, nearly upright shoot made four revolutions, moving against the sun, very regularly at an average rate of 3 hrs. 26 m. The shoots, however, sometimes stood still. It is considered a greenhouse plant; but when kept there, the petioles took several days to clasp a stick: in the hothouse, a stick was clasped in 7 hrs. In the greenhouse, a petiole was not affected by a loop of string, suspended during several days and weighing 2½ grains; but in the hothouse, one was made to curve by a loop weighing 1.64 gr.; and, on the removal of the string, it became straight again.

The flexible petiole of a half or a quarter grown leaf which has clasped an object for three or four days increases much in thickness, and after several weeks becomes so wonderfully hard and rigid that it can hardly be

Solanum jasminoides, with one of its petioles clasping a stick.

removed from its support. On comparing a thin transverse slice of such a petiole with one from an older leaf growing close beneath, which had not clasped anything, its diameter was found to be fully doubled, and its structure greatly changed. In two other petioles similarly compared, and here represented, the increase in diameter was not quite so great. In the section of the petiole in its ordinary state (A), we see a semilunar band of cellular tissue (not well shown in the woodcut) differing slightly in appearance from that outside it and including three closely approximate groups of dark vessels. Near the upper surface of the petiole, beneath two exterior ridges, there are two other small circular groups of vessels. In the section of the petiole (B) which had clasped during several weeks a stick, the two exterior ridges have become much less prominent, and the two groups of woody vessels beneath them much increased in diameter. The semilunar band has been converted into a complete ring of very hard, white, woody tissue, with lines radiating from the centre. ... This petiole after clasping the stick had actually become thicker than the stem from which it arose; and this was chiefly due to the increased thickness of the ring of wood. ... It is a singular morphological fact that the petiole should thus acquire a structure almost identically the same with that of the axis; and it is a still more singular physiological fact that so great a change should have been induced by the mere act of clasping a support.

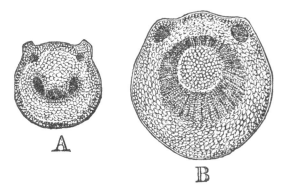

Solanum jasminoides. A. Section of a petiole in its ordinary state.
B. Section of a petiole some weeks after it had clasped a stick.

Spiranthes cernua. Hand-colored engraving drawn by Walter Hood Fitch,
from *The Botanical* Magazine 87: 5277.

Spiranthes
LADIES TRESSES

———— ...... ————

ORCHIDACEAE—ORCHID FAMILY

ORCHIDS, FORMS OF FLOWERS, POLLINATION

Spiranthes, with the common name ladies tresses describing its attractive attire, is a cosmopolitan orchid genus of about forty species, distinguished by spiraled spikes of flowers fertilized by moths and bees. Darwin obtained plants of *Spiranthes autumnalis* (= *S. spiralis*) from his correspondent Alexander More on the Isle of Wight, and also found them while on vacation at Torquay. In his orchid book, Darwin described the long, flat "boat-formed" structure (*viscidium*) attached at the base of the pollinia (pollen packets), bearing adhesive fluid that guaranteed strong attachment to a pollinator's proboscis (or even a faux proboscis—Darwin and More experimented with blades of grass and needles, poking them into flowers to trigger the "curious contrivance" of movement and transfer of the pollinia). He saw a bee with five such "boats" stuck to its proboscis, carrying them away to another flower for pollination.

Darwin's delight and wonder at the elaborate pollination mechanism of *Spiranthes* is evident in his meticulous description, from the fit of an insect's proboscis in the flower to how the flowers transition from female to male up the spike, termed "protandry" today. The flowers are functionally male at first, presenting pollinia but not allowing access to the stigma. They open further as they age— hence, the older, lower, flowers in a spike become pollinated first since insects tend to "alight at the bottom and crawl up the spike," as Darwin observed, depositing pollen from the top-most flowers of the last spike they visited and, making their way up, picking up a fresh supply for the next plant.

Eager to see how general his findings were, Darwin asked botanist Asa Gray to examine the North American species *Spiranthes cernua* and *S. gracilis*: "I enclose sketch about *Spiranthes*; if you will observe your species, I should be infinitely obliged." Gray later confirmed the observations: "The difference between the older flowers and those first opened is striking. The latter presents the disc—the former the stigma."[137]

[
From: *The Various Contrivances by which Orchids are Fertilised by Insects* (2nd ed., 1877)
]

Spiranthes autumnalis.—This Orchid with its pretty name of Ladies'-tresses, presents some interesting peculiarities. The rostellum is a long, thin, flat projection, joined by sloping shoulders to the summit of the stigma. In the middle of the rostellum a narrow vertical brown object (fig. C) may be seen, bordered and covered by transparent membrane. This brown object I will call "the boat-formed disc." It forms the middle portion of the posterior surface of the rostellum and consists of a narrow strip of the exterior membrane in a modified condition. When removed from its attachment, its summit (fig. E) is seen to be pointed, with the lower end rounded; it is slightly bowed, so as altogether to resemble a boat or canoe. ...

The stigma lies beneath the rostellum and projects with a sloping surface, as may be seen at B in the side-view: its lower margin is rounded and fringed with hairs. On each side a membrane (*cl*, B) extends from the edges of the stigma to the filament of the anther, thus forming a membranous cup or clinandrum, in which the lower ends of the pollen-masses lie safely protected. ...

The tubular flowers are elegantly arranged in a spire round the spike, and project from it horizontally (fig. A). The labellum is channelled down the middle, and is furnished with a reflexed and fringed lip, on which bees alight; its basal internal angles are produced into two globular processes, which secrete an abundance of nectar. The nectar is collected (*n*, fig. B) in

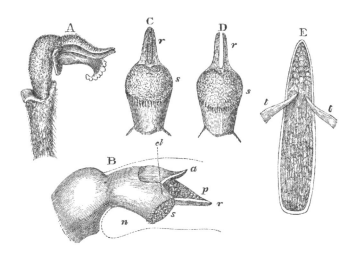

Spiranthes autumnalis. A. Side view of flower in its natural position, with the two lower sepals alone removed. The labellum can be recognised by its fringed and reflexed lip. B. Side view of a mature flower, with all the sepals and petals removed. The position of the labellum (which has moved from the rostellum) and the upper sepal is shown by the dotted lines. C. Front view of the stigma, and of the rostellum with its embedded central disc. D. Front view of the stigma and of the rostellum after the viscid disc has been removed. E. Viscid disc, removed from the rostellum, greatly magnified, viewed posteriorly, and with the attached elastic threads of the pollen-masses; the pollen-grains have been removed from the threads.

a. anther. *p.* pollen-masses. *t.* threads of the pollen-masses. *cl.* margin of clinandrum. *r.* rostellum. *s.* stigma. *n.* nectar-receptacle.

a small receptacle in the lower part of the labellum. Owing to the protuberance of the inferior margin of the stigma and of the two lateral inflexed nectaries, the orifice into the nectar-receptacle is much contracted. When the flower first opens, the receptacle contains nectar, and at this period, the front of the rostellum, which is slightly furrowed, lies close to the channelled labellum; consequently a passage is left, but so narrow that only a fine bristle can be passed down it. In a day or two, the column moves a little farther from the labellum, and a wider passage is left for insects to deposit pollen on the stigmatic surface. On this slight movement of the column the fertilisation of the flower absolutely depends. ...

We thus see how beautifully everything is contrived that the pollinia should be withdrawn by insects visiting the flowers. They are already attached to the disc by their threads, and, from the early withering of the anther-cells, they hang loosely suspended but protected within the clinandrum. The touch of the proboscis causes the rostellum to split in front and behind, and frees the long, narrow, boat-formed disc, which is filled with extremely viscid matter, and is sure to adhere longitudinally to the proboscis. When the bee flies away, so surely will it carry away the pollinia. As the pollinia are attached parallel to the disc, they adhere parallel to the proboscis. When the flower first opens and is best adapted for the removal of the pollinia, the labellum lies so close to the rostellum that the pollinia attached to the proboscis of an insect cannot possibly be forced into the passage so as to reach the stigma; they would be either upturned or broken off: but we have seen that after two or three days, the column becomes more reflexed and moves from the labellum,—a wider passage being thus left. When I inserted the pollinia attached to a fine bristle into the nectar-receptacle of a flower in this condition (n, fig. B), it was pretty to see how surely the sheets of pollen were left adhering to the viscid stigma. ...

Hence in *Spiranthes*, recently expanded flower, which has its pollinia in the best state for removal, cannot be fertilised; and mature flowers will be fertilised by pollen from younger flowers, borne, as we shall presently see, on a separate plant. In conformity with this fact, the stigmatic surfaces of the older flowers are far more viscid than those of the younger flowers. Nevertheless, a flower which in its early state had not been visited by insects would not necessarily, in its later and more expanded condition, have its pollen wasted: for insects, in inserting and withdrawing their proboscides, bow them forwards or upwards, and would thus often strike the furrow in the rostellum. I imitated this action with a bristle, and often succeeded in withdrawing the pollinia from old flowers. I was led to make this trial from having at first chosen old flowers for examination; and on passing a bristle, or fine culm of grass, straight down into the nectary, the pollinia were never withdrawn; but when it was bowed forward, I succeeded. Flowers which have not had their pollinia removed can be fertilised as easily as those which have lost them; and I have seen not a few cases of flowers with their pollinia still in place, with sheets of pollen on their stigmas.

At Torquay I watched for about half an hour a number of these flowers growing together, and saw three humble-bees of two kinds visit them. ... The next day I watched the same flowers for a quarter of an hour and caught another humble-bee at work; one perfect pollinium and four boat-formed discs adhered to its proboscis, one on the top of the other, showing how exactly the same part of the rostellum had each time been touched.

The bees always alighted at the bottom of the spike, and, crawling spirally up it, sucked one flower after the other. I believe humble-bees generally act in this manner when visiting a dense spike of flowers, as it is the most convenient method; on the same principle that a woodpecker always climbs up a tree in search of insects. This seems an insignificant observation; but see the result. In the early morning, when the bee starts on her rounds, let us suppose that she alighted on the summit of a spike; she would certainly extract the pollinia from the uppermost and last opened flowers; but when visiting the next suc-ceeding flower, of which the column in all probability would not as yet have moved from the labellum (for this is slowly and very gradually effected), the pollen-masses would be brushed off her proboscis and wasted. But nature suffers no such waste. The bee goes first to the lowest flower, and, crawling spirally up the spike, effects nothing on the first spike which she visits till she reaches the upper flowers, and then she withdraws the pollinia. She soon flies to another plant, and, alighting on the lowest and oldest flower, into which a wide passage will have been formed from the greater reflexion of the column, the pollinia strike the protuberant stigma. If the stigma of the lowest flower has already been fully fertilised, little or no pollen will be left on its dried surface; but on the next succeeding flower, of which the stigma is adhesive, large sheets of pollen will be left. Then as soon as the bee arrives near the summit of the spike she will withdraw fresh pollinia, will fly to the lower flowers on another plant, and fertilise them; and thus, as she goes her rounds and adds to her store of honey, she continually fertilises fresh flowers and perpetuates the race of our autumnal *Spiranthes*, which will yield honey to future generations of bees.

Trifolium pratense   Red Clover.

Trifolium pratense. Watercolor by Elizabeth Wharton, *British Flowers*.

Trifolium

CLOVER

......

FABACEAE—PEA FAMILY

CROSS AND SELF-FERTILIZATION, PLANT MOVEMENT

Clovers of the world are commonly trifoliate, bearing compound leaves of three leaflets (hence the Latin *trifolium*) and heads of small densely clustered flowers of red, purple, white, or yellow. A common and seemingly unremarkable genus of field and lawn, several of the 300 or more *Trifolium* species are grown worldwide for fodder, hay, and silage, as well as pasture improvement thanks to their nitrogen-fixing roots. Their nectar-rich flowers have also made them a favorite of beekeepers. Darwin encouraged clovers in his meadows and gardens for another reason: he saw in them evolutionary dramas.

It started with cross-pollination studies. He was sure that bees, especially "humble-bees" (bumblebees) were vital to the cross-pollination of many flowering plants, but he needed data to back up his hunch. Turning to patches of red clover (*T. pratense*) in the meadow just behind his house, he performed a classic exclusion experiment, netting some patches to keep bees out while leaving others open, then counting the seeds produced by flowers in each group. It was, as he excitedly reported to Joseph Hooker, a success. "100 Head of *Trifolium pratense* protected from Bees did not produce *one single seed*; another Hundred visited by Humble-bees produced nearly 3000 seeds!"[138] It got him thinking: if bees declined in the area, what would happen to the flowers? Further, what species might affect the bees? And what species might affect the species that affect the bees? This led to the classic passage in *Origin* quoted in this chapter, which is, perhaps, the first description of a chain of ecological interconnectedness, connecting cats in an area to its flower populations through the intermediaries of mice and bumblebees.[139]

[From: *On the Origin of Species* (1859)]

I am tempted to give one more instance showing how plants and animals, most remote in the scale of nature, are bound together by a web of complex relations. ... From experiments which I have tried, I have found that the visits of bees, if not indispensable, are at least highly beneficial to the fertilisation of our clovers; but humble-bees alone visit the common red clover (*Trifolium pratense*), as other bees cannot reach the nectar. Hence I have very little doubt, that if the whole genus of humble-bees became extinct or very rare in England, the heartsease and red clover would become very rare, or wholly disappear. The number of humble-bees in any district depends in a great degree on the number of field-mice, which destroy their combs and nests; and Mr. H. Newman, who has long attended to the habits of humble-bees, believes that "more than two thirds of them are thus destroyed all over England." Now the number of mice is largely dependent, as every one knows, on the number of cats; and Mr. Newman says, "Near villages and small towns I have found the nests of humble-bees more numerous than elsewhere, which I attribute to the number of cats that destroy the mice." Hence it is quite credible that the presence of a feline animal in large numbers in a district might determine, through the intervention first of mice and then of bees, the frequency of certain flowers in that district! ...

Let us now turn to the nectar-feeding insects in our imaginary case: we may suppose the plant of which we have been slowly increasing the nectar by continued selection, to be a common plant; and that certain insects depended in main part on its nectar for food. I could give many facts, show-ing how anxious bees are to save time; for instance, their habit of cutting holes and sucking the nectar at the bases of certain flowers, which they can, with a very little more trouble, enter by the mouth. Bearing such facts in mind, I can see no reason to doubt that an accidental deviation in the size and form of the body, or in the curvature and length of the proboscis, &c., far too slight to be appreciated by us, might profit a bee or other insect, so that an individual so characterised would be able to obtain its food more quickly, and so have a better chance of living and leaving descendants. Its descendants would probably inherit a tendency to a similar slight devia-tion of structure. The tubes of the corollas of the common red and incar-nate clovers (*Trifolium pratense* and *incarnatum*) do not on a hasty glance appear to differ in length; yet the hive-bee can easily suck the nectar out of the incarnate clover, but not out of the common red clover, which is visited by humble-bees alone; so that whole fields of the red clover offer in vain an abundant supply of precious nectar to the hive-bee. Thus it might be a great advantage to the hive-bee to have a slightly longer or differently con-structed proboscis. On the other hand, I have found by experiment that the fertility of clover greatly depends on bees visiting and moving parts of the corolla, so as to push the pollen on to the stigmatic surface. Hence, again, if humble-bees were to become rare in any country, it might be a great advan-tage to the red clover to have a shorter or more deeply divided tube to its corolla, so that the hive-bee could visit its flowers. Thus I can understand how a flower and a bee might slowly become, either simultaneously or one after the other, modified and adapted in the most perfect manner to each other, by the continued preservation of individuals presenting mutual and slightly favourable deviations of structure.[140]

Things were a bit more complicated than Darwin realized, however. He believed that bumblebees alone pollinated red clover, as a honey-bee, with its shorter proboscis, could not reach the nectar.

$$\left[\begin{array}{l}\text{From: } \textit{The Effects of Cross and Self Fertilisation in}\\ \quad\quad \textit{the Vegetable Kingdom (2nd ed., 1878)}\end{array}\right]$$

Trifolium repens (Leguminosae).—Several plants were protected from insects, and the seeds from ten flower-heads on these plants, and from ten heads on other plants growing outside the net (which I saw visited by bees), were counted; and the seeds from the latter plants were very nearly ten times as numerous as those from the protected plants. The experiment was repeated on the following year; and twenty protected heads now yielded only a single aborted seed, whilst twenty heads on the plants outside the net (which I saw visited by bees) yielded 2290 seeds, as calculated by weighing all the seed, and counting the number in a weight of two grains.

T. pratense.—One hundred flower-heads on plants protected by a net did not produce a single seed, whilst 100 heads on plants growing outside, which were visited by bees, yielded 68 grains weight of seeds; and as eighty seeds weighed two grains, the 100 heads must have yielded 2,720 seeds. I have often watched this plant, and have never seen hive-bees sucking the flowers, except from the outside through holes bitten by humble-bees, or deep down between the flowers, as if in search of some secretion from the calyx ... It is at least certain that humble-bees are the chief fertilisers of the common red clover.

A vicar friend, Charles Hardy, corrected Darwin on this point, relating evidence of a red clover field teeming with honeybees (though he acknowledged that it was second-growth, after mowing, and the flowers seemed smaller).[141] Darwin added Hardy's observation to the third edition of *Origin*, but this also led him down a rabbit-hole (all good science has them). Watching honeybees visiting red clover, he noticed that some were nectar robbers, getting nectar from the base of the flower through holes chewed through the corolla, while others went into the flower in the usual way. He suddenly thought these behaviors could reflect separate foraging castes of bees, with long and short proboscas. As he was on holiday in Southampton with his family when this occurred to him, he couldn't find a clover field to test the idea, so he wrote an urgent letter to his neighbor John Lubbock, asking him to look out for these modes of foraging in his local bees, urging, "Now if you see this, do for Heaven sake catch me some of each and put in spirits and *keep them separate*." But by the next day, he'd succeeded in finding a field and realized the idea was wrong, writing as follow up, "I beg a million pardons … it is all an illusion (but almost excusable) about the Bees. I do so hope that you have not wasted any time for my stupid blunder." He closed with tongue-in-cheek self-flagellation: "I hate myself I hate clover and I hate Bees."[142]

Darwin's later interests in *Trifolium* extended to plant movement, with analyses of the nyctitropic (sleep) movement of leaves and circular (circumnutation) motion of cotyledons and flower stems. One of the most remarkable to him was the geotropic (gravity-sensitive) motion of *T. subterraneum*, the burrowing clover. This species is unique among clovers in that it buries its flowers after fertilization—the seeds develop underground, a phenomenon called geocarpy. "There seems much odd about whole case," Darwin commented to his son Francis as he tried to figure out how the plant buried its own seeds.[143] He eventually succeeded and reported how circumnutation of *T. subterraneum* peduncles continues even after the flower heads touch the ground, slowing "planting" their own seeds.

[**From: *The Power of Movement in Plants* (1880)**]

The Burying of Seed-capsules: Trifolium subterraneum.—The flower-heads of this plant are remarkable from producing only 3 or 4 perfect flowers, which are situated exteriorly. All the other many flowers abort, and are modified into rigid points, with a bundle of vessels running up their centres. After a time 5 long, elastic, claw-like projections, which represent the divisions of the calyx, are developed on their summits. As soon as the perfect flowers wither they bend downwards, supposing the peduncle to stand upright, and they then closely surround its upper part. This movement is due to epinasty, as is likewise the case with the flowers of *T. repens*. The imperfect central flowers ultimately follow, one after the other, the same course. Whilst the perfect flowers are thus bending down, the whole peduncle curves downwards and increases much in length, until the flower-head reaches the ground. ... In whatever position the branches may be placed, the upper part of the peduncle at first bends vertically upwards through heliotropism; but as soon as the flowers begin to wither the downward curvature of the whole peduncle commences. As this latter movement occurred in complete darkness, and with peduncles arising from upright and from dependent branches, it cannot be due to apheliotropism or to epinasty, but must be attributed to geotropism.

After the heads have buried themselves, the central aborted flowers increase considerably in length and rigidity, and become bleached. They gradually curve, one after the other, upwards or towards the peduncle, in the same manner as did the perfect flowers at first. In thus moving, the long claws on their summits carry with them some earth. Hence a flower-head which has been buried for a sufficient time, forms a rather large ball, consisting of the aborted flowers, separated from one another by earth, and surrounding the little pods (the product of the perfect flowers) which lie close round the upper part of the peduncle. The calyces of the perfect and imperfect flowers are clothed with simple and multicellular hairs, which have the power of absorption; for when placed in a weak solution of carbonate of ammonia (2 gr. to 1 oz. of water) their protoplasmic contents immediately became aggregated and afterwards displayed the usual slow movements. This clover generally grows in dry soil, but whether the power of absorption by the hairs on the buried flower-heads is of any importance to them we do not know. Only a few of the flower-heads, which from their position are not able to reach the ground and bury themselves, yield seeds; whereas the buried ones never failed, as far as we observed, to produce as many seeds as there had been perfect flowers.

We will now consider the movements of the peduncle whilst curving down to the ground. We have seen that an upright young flower-head circumnutated conspicuously; and that this movement continued after the peduncle had begun to bend downwards. The same peduncle was observed when inclined at an angle of 19° above the horizon, and it circumnutated during two days. ... During the first day the peduncle clearly circumnutated, for it moved 4 times down and 3 times up; and on each succeeding day, as it sank downwards, the same movement continued. ...

Any one who will observe a flower-head burying itself, will be convinced that the rocking movement, due to the continued circumnutation of the peduncle, plays an important part in the act. Considering that the flower-heads are very light, that the peduncles are long, thin, and flexible, and that they arise from flexible branches, it is incredible that an object as blunt as one of these flower-heads could penetrate the ground by means of the growing force of the peduncle, unless it were aided by the rocking movement. After a flower-head has penetrated the ground to a small depth, another and efficient agency comes into play; the central rigid aborted flowers, each terminating in five long claws, curve up towards the peduncle; and in doing so can hardly fail to drag the head down to a greater depth, aided as this action is by the circumnutating movement, which continues after the flower-head has completely buried itself. The aborted flowers thus act something like the hands of the mole, which force the earth backwards and the body forwards.

AGRIUIOLA maxima
odorata. Boerh.

Jndian Crefs.

Tropaeolum majus. Water and bodycolor on vellum by English School artist,
Album of Garden Flowers.

```
                    Tropaeolum
           NASTURTIUM
           ────── ...... ──────
      TROPAEOLACEAE—NASTURTIUM FAMILY
```

CLIMBING PLANTS, MOVEMENT

The Nasturtium family has but a single genus, *Tropaeolum*, comprising some eighty species widespread in Central and South America. The popular garden nasturtium (*T. majus*), canary bird vine (*T. peregrinum*), and flame nasturtium (*T. speciosum*) are prized by gardeners worldwide for both their culinary use—peppery in flavor, like watercress in the Mustard family (genus *Nasturtium*, not coincidentally)—and vividly colored flowers. Those striking colors inspired a sanguinary reference by Linnaeus in bestowing the name *Tropaeolum*, which is derived from the Greek *tropaion*, "trophy." This references the ancient Roman custom of the trophy-pole, on which vanquished enemies' armor and weapons were hung. The round leaves of the type species *T. majus* were imagined to resemble shields and its flowers blood-stained helmets.

Had Linnaeus waited another decade before naming *Tropaeolum*, an observation made by his nineteen-year-old daughter Elisabet Christina might have inspired a more interesting moniker. In the fading light of dusk one evening, she noticed that the flowers suddenly gleamed, almost seeming to emit flashes or sparks of light. Her observations were soon published,[144] touching off intense scientific speculation and poetic inspiration over what became known as the "Elizabeth Linnaeus Phenomenon." Erasmus Darwin wrote in *The Loves of the Plants* how in the twilight, "A faint glory trembles" around *Tropaeolum*, over whose "fair form the electric lustre plays," and in the notes cited Elisabet Christina's discovery of what many took to be an electrical phenomenon.[145]

The Romantic poets were inspired in turn. Coleridge, for example, drew on *The Loves of the Plants* in penning the lines: "'Tis said, in Summer's evening hour / Flashes the golden-colour'd flower / A fair electric flame."[146] More recent studies confirm an explanation first suggested by the German polymath Johann Wolfgang von Goethe: the scintillating gleam is more optical illusion than botanical electricity, created when the brightly colored flowers are viewed obliquely against a leafy green background in low light.[147]

Charles Darwin embraced a different kind of botanical "electricity," pertaining more to impulses, analogous to the animal nervous system, thought to be behind the touch-sensitivity and rapid movement of some plants. If he knew of Elisabet Christina Linnaea and the case of the curiously scintillating *Tropaeolum*, it is not clear that he ever commented on it, his interest in nasturtiums falling squarely within the realms of pollination, inheritance, and movement. In *Cross and Self Fertilisation*, he noted that nasturtiums are "manifestly adapted for cross-fertilisation by insects" due to their protandrous flowers, with stamens developing before the pistils so that insects bring pollen from a younger (male) flower to an older (female) one. His experiments resulted in a batch of crossed plants producing 243 seeds, while the same number of self-fertilized plants yielded only 155 seeds—consistent with his overarching hypothesis of the benefits of outcrossing. Further research on several species and varieties involved movement, documenting how young revolving internodes and petioles enable these leaf-climbers to ascend supporting objects (as described in *Climbing Plants*), and the heliotropic (turning toward sunlight) and nyctitropic (nocturnal "sleep") movement of leaves (as described in *Movement*).

From: *The Movements and Habits of Climbing Plants*
(2nd ed., 1875)

Tropaeolum tricolorum, var. *grandiflorum.*—The flexible shoots, which first
rise from the tubers, are as thin as fine twine. One such shoot revolved in
a course opposed to the sun, at an average rate, judging from three revolutions, of 1 hr. 23 m.; but no doubt the direction of the revolving movement is variable. When the plants have grown tall and are branched, all
the many lateral shoots revolve. The stem, whilst young, twines regularly
round a thin vertical stick, and in one case I counted eight spiral turns in the
same direction; but when grown older, the stem often runs straight up for a
space, and, being arrested by the clasping petioles, makes one or two spires
in a reversed direction. Until the plant grows to a height of two or three
feet, requiring about a month from the time when the first shoot appears
above ground, no true leaves are produced, but, in their place, filaments
coloured like the stem. The extremities of these filaments are pointed, a
little flattened, and furrowed on the upper surface. They never become
developed into leaves. As the plant grows in height new filaments are produced with slightly enlarged tips; then others, bearing on each side of the
enlarged medial tip a rudimentary segment of a leaf; soon other segments
appear, and at last a perfect leaf is formed, with seven deep segments. So
that on the same plant we may see every step, from tendril-like clasping
filaments to perfect leaves with clasping petioles. After the plant has grown
to a considerable height and is secured to its support by the petioles of the
true leaves, the clasping filaments on the lower part of the stem wither and
drop off; so that they perform only a temporary service.

These filaments or rudimentary leaves, as well as the petioles of the
perfect leaves, whilst young, are highly sensitive on all sides to a touch.
The slightest rub caused them to curve towards the rubbed side in about
three minutes, and one bent itself into a ring in six minutes; they subsequently became straight. When, however, they have once completely
clasped a stick, if this is removed, they do not straighten themselves. The

most remarkable fact, and one which I have observed in no other species of the genus, is that the filaments and the petioles of the young leaves, if they catch no object, after standing for some days in their original position, spontaneously and slowly oscillate a little from side to side, and then move towards the stem and clasp it. They likewise often become, after a time, in some degree spirally contracted. They therefore fully deserve to be called tendrils, as they are used for climbing, are sensitive to a touch, move spontaneously, and ultimately contract into a spire, though an imperfect one. The present species would have been classed amongst the tendril-bearers, had not these characters been confined to early youth. During maturity it is a true leaf-climber. ...

[From: *The Power of Movement in Plants* (1880)]

Tropaeolum majus (cultivated var.) —Several plants in pots stood in the greenhouse, and the blades of the leaves which faced the front-lights were during the day highly inclined and at night vertical; whilst the leaves on the back of the pots, though of course illuminated through the roof, did not become vertical at night. We thought, at first, that this difference in their positions was in some manner due to heliotropism, for the leaves are highly heliotropic. The true explanation, however, is that unless they are well illuminated during at least a part of the day they do not sleep at night; and a little difference in the degree of illumination determines whether or not they shall become vertical at night.

A large pot with several plants was brought on the morning of Sept. 3rd out of the greenhouse and placed before a north-east window, in the same position as before with respect to the light, as far as that was possible. On the front of the plants, 24 leaves were marked with thread, some of which had their blades horizontal, but the greater number were inclined at about 45°, beneath the horizon; at night all these, without exception, became vertical. Early on the following morning (4th) they reassumed their former positions, and at night again became vertical. On the 5th the shutters were opened at 6.15 A.M., and by 8.18 A.M., after the leaves had been illuminated for 2 h. 3 m. and had acquired their diurnal position, they were placed in a dark cupboard. They were looked at twice during the day and thrice in the evening, the last time at 10.30 P.M., and not one had become vertical. At 8 A.M. on the following morning (6th) they still retained the same diurnal position and were now replaced before the north-east window. At night all the leaves which had faced the light had their petioles curved and their blades vertical; whereas none of the leaves on the back of the plants, although they had been moderately illuminated by the diffused light of the room, were vertical. They were now at night placed in the same dark cupboard; at 9 A.M. on the next morning (7th) all those which had been asleep had reassumed their diurnal position. The pot was then placed for 3 h. in the sunshine, so as to stimulate the plants; at noon they were placed before the same north-east window, and at night the leaves slept in the usual manner and awoke on the following morning. At noon on this day (8th) the plants, after having been left before the north-east window for 5 h. 45 m. and thus illuminated (though not brightly, as the sky was cloudy during the whole time), were replaced in the dark cupboard, and at 3 P.M. the position of the leaves was very little, if at all, altered, so that they are not quickly affected by darkness; but by 10.15 P.M. all the leaves which had faced the north-east sky during the 5 h. 45 m. of illumination stood vertical, whereas those on the back of the plant retained their diurnal position. On the following morning (9th) the leaves awoke as on the two former occasions in the dark, and they were kept in the dark during the whole day; at night a very few of them became vertical, and this was the one instance in which we observed any inherited tendency or habit in this plant to sleep at the proper time. That it was real sleep was shown by these same leaves reassuming their diurnal position on the following morning (10th) whilst still kept in the dark.

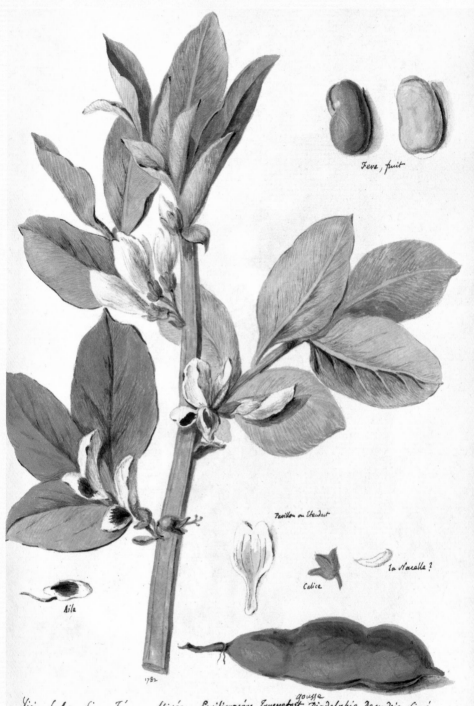

Vicia faba. Watercolor on paper by Charles Germain de Saint-Aubin, *Receuil De Plantes Copiées D 'Aprés Nature Par De Saint Aubin Dessinateur Du Roy Louis XV.*

Vicia faba

BROAD or FAVA BEANS

—— ...... ——

FABACEAE—PEA FAMILY

PLANT MOVEMENT

The broad bean or fava bean, a culinary staple since antiquity, is, botanically speaking, a species of vetch, belonging to a group of nearly 150, mainly annual, climbing legumes in the genus *Vicia*. The family name Fabaceae is a vestige of botanical history—during a brief time when these beans were placed in genus *Faba* (Latin for "bean") they were selected as the taxonomic representative of legumes—making *Faba faba*, the "bean of beans," the representative species—and so the genus was built into the family name. The genus name *Faba* has long since been defunct, replaced with *Vicia*, but it lives on in the higher classification. That's the sort of relict, so to speak, that Darwin found instructive as an analog of evolutionary history—like the way oddities of spelling or silent letters can reveal the origin and evolution of certain words. But this nomenclatural example didn't exist in Darwin's day, when legumes were variously known by the more descriptive names "Leguminosae" or "Papilionaceae." His interest in them had more to do with pollination, movement, and physiology.

In the course of his investigations into the fertility of cultivated beans, Darwin found that flowers of fava beans have to be cross-pollinated by insects to ensure proper seed set. He demonstrated this in straightforward experiments, such as the one in which he netted one group of fava beans in his garden to exclude bees and left others open, harvesting 135 beans from seventeen "open pollinated" plants, compared to a measly 40 beans from a like number of covered ones. But the study of fava bean pollination was minor compared to his years-long focus on the many forms of movement in this species, including circular-motion circumnutation of leaves, where he observed that "both the whole leaf and the terminal leaflets undergo a well-marked daily periodical movement, rising in the evening and falling during the latter part of the night." Even more elaborate were studies of seedlings, testing circumnutation of the epicotyl (the stem region above the cotyledon of a seedling), radicle (the embryonic root), and hypocotyl (the region between the cotyledon and radicle). Readily available and easy to germinate, fava beans were ideal for such research.

Darwin and his son Francis carefully traced the movement of fava bean radicles as they grew, and, by adhering tiny card rectangles in different locations near the apex, made the surprising discovery that the radicles are touch-sensitive—they tend to curve away from the side touched, just the opposite of the reaction observed with tendrils. These and other experiments, such as the reaction of radicle tips to light and gravity, led the father-son team to declare that "there is no structure in plants more wonderful, as far as its functions are concerned, than the tip of the radicle." But how can cells at the tip of this structure influence cell growth further up the elongating root stem? They marveled that the tip acts somewhat like a brain. In the very last sentence of *Movement* they developed this intriguing analogy: "It is hardly an exaggeration to say that the tip of the radicle thus endowed [with sensitivity] and having the power of directing the movements of the adjoining parts, acts like the brain of one of the lower animals; the brain being seated within the anterior end of the body, receiving impressions from the sense-organs, and directing the several movements."[148] Later known as Darwin's "root-brain hypothesis" and largely discounted in the twentieth century, this idea has re-emerged in recent years in the form of a thought-provoking "phytoneurobiological" model for understanding plant growth and physiology.[149] When next you sit down to a hearty dish of fava beans, raise a fork to this humble bean that has not only played a leading role in the development of the field of plant physiology but continues to instruct as a staple, so to speak, of high-school and college botany classes today.

[**From: *The Power of Movement in Plants* (1880)**]

In order to see how the radicles of seedlings would pass over stones, roots, and other obstacles, which they must incessantly encounter in the soil, germinating beans (*Vicia faba*) were so placed that the tips of the radicles came into contact, almost rectangularly or at a high angle, with underlying plates of glass. In other cases, the beans were turned about whilst their radicles were growing, so that they descended nearly vertically on their own smooth, almost flat, broad upper surfaces. The delicate root-cap, when it first touched any directly opposing surface, was a little flattened transversely; the flattening soon became oblique, and in a few hours quite disappeared, the apex now pointing at right angles, or at nearly right angles, to its former course. The radicle then seemed to glide in its new direction over the surface which had opposed it, pressing on it with very little force. How far such abrupt changes in its former course are aided by the circumnutation of the tip must be left doubtful. Thin slips of wood were cemented on more or less steeply inclined glass-plates, at right angles to the radicles which were gliding down them. Straight lines had been painted along the growing terminal part of some of these radicles, before they met the opposing slip of wood; and the lines became sensibly curved in 2 h. after the apex had come into contact with the slips. In one case of a radicle, which was growing rather slowly, the root-cap, after encountering a rough slip of wood at right angles, was at first slightly flattened transversely: after an interval of 2 h. 30 m. the flattening became oblique; and after an additional 3 hours the flattening had wholly disappeared, and the apex now pointed at right angles to its former course. It then continued to grow in its new direction alongside the slip of wood, until it came to the end of it, round which it bent rectangularly. Soon afterwards when coming to the edge of the plate of glass, it was again bent at a large angle, and descended perpendicularly into the damp sand. ...

... An object which yields with the greatest ease will deflect a radicle: thus, as we have seen, when the apex of the radicle of the bean encountered the polished surface of extremely thin tin-foil laid on soft sand, no impression was left on it, yet the radicle became deflected at right angles. A second explanation occurred to us, namely, that even the gentlest pressure might check the growth of the apex, and in this case growth could continue only on one side, and thus the radicle would assume a rectangular form; but this view leaves wholly unexplained the curvature of the upper part, extending for a length of 8–10 mm.

We were therefore led to suspect that the apex was sensitive to contact, and that an effect was transmitted from it to the upper part of the radicle, which was thus excited to bend away from the touching object. As a little loop of fine thread hung on a tendril or on the petiole of a leaf-climbing plant, causes it to bend, we thought that any small hard object affixed to the tip of a radicle, freely suspended and growing in damp air, might cause it to bend, if it were sensitive, and yet would not offer any mechanical resistance to its growth. Full details will be given of the experiments which were tried, as the result proved remarkable. The fact of the apex of a radicle being sensitive to contact has never been observed, though, as we shall hereafter see, Sachs discovered that the radicle a little above the apex is sensitive and bends like a tendril *towards* the touching object. But when one side of the apex is pressed by any object, the growing part bends *away* from the object; and this seems a beautiful adaptation for avoiding obstacles in the soil, and, as we shall see, for following the lines of least resistance. Many organs, when touched, bend in one fixed direction, such as the stamens of *Berberis*, the lobes of *Dionaea*, etc.; and many organs, such as tendrils, whether modified leaves or flower-peduncles, and some few stems, bend towards a touching object; but no case, we believe, is known of an organ bending away from a touching object.

Sensitiveness of the Apex of the Radicle of Vicia faba.—Common beans, after being soaked in water for 24 h., were pinned with the hilum downwards (in the manner followed by Sachs), inside the cork lids of glass-vessels, which were half filled with water; the sides and the cork were well moistened, and light was excluded. As soon as the beans had protruded radicles, some to a length of less than a tenth of an inch, and others to a length of several tenths, little squares or oblongs of card were affixed to the short sloping sides of their conical tips. The squares therefore adhered obliquely with reference to the longitudinal axis of the radicle; and this is a very necessary precaution, for if the bits of card accidentally became displaced, or were drawn by the viscid matter employed so as to adhere parallel to the side of the radicle, although only a little way above the conical apex, the radicle did not bend in the peculiar manner which we are here considering. Squares of about the 1/20th of an inch (i.e. about 1½ mm.), or oblong bits of nearly the same size, were found to be the most convenient and effective. We employed at first ordinary thin card, such as visiting cards, or bits of very thin glass, and various other objects; but afterwards sand-paper was chiefly employed, for it was almost as stiff as thin card, and the roughened surface favoured its adhesion. At first we generally used very thick gum-water; and this of course, under the circumstances, never dried in the least; on the contrary, it sometimes seemed to absorb vapour, so that the bits of card became separated by a layer of fluid from the tip. When there was no such absorption and the card was not displaced, it acted well and caused the radicle to bend to the opposite side. I should state that thick gum-water by itself induces no action. In most cases the bits of card were touched with an extremely small quantity of a solution of shellac in spirits of wine, which had been left to evaporate until it was thick; it then set hard in a few seconds and fixed the bits of card well. When small drops of the shellac were placed on the tips without any card, they set into hard little beads, and these acted like any other hard object, causing the radicles to bend to the opposite side. ...

... As the chief curvature of the radicle is at a little distance from the apex, and as the extreme terminal and basal portions are nearly straight, it is possible to estimate in a rough manner the amount of curvature by an angle; and when it is said that the radicle became deflected at any angle from the perpendicular, this implies that the apex was turned upwards by so many degrees from the downward direction which it would natu- rally have followed, and to the side opposite to that to which the card was affixed. That the reader may have a clear idea of the kind of movement excited by the bits of attached card, we append here accurate sketches of three germinating beans thus treated and selected out of several specimens to show the gradations in the degrees of curvature.

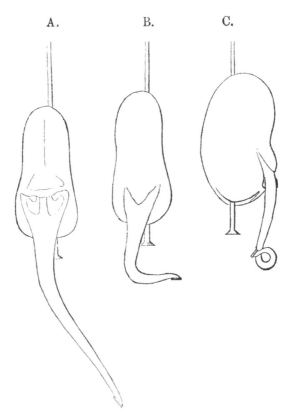

Vicia faba: A, radicle beginning to bend from the attached little square of card; B, bent at a rectangle; C, bent into a circle or loop, with the tip beginning to bend downwards through the action of geotropism.

Vinca minor. Watercolor by artist at French School, *An Album of Flowers.*

Vinca

PERIWINKLE

——— ...... ———

APOCYNACEAE—DOGBANE FAMILY

Periwinkle is a common name for plants in two related genera in this family—*Vinca*, a small genus of seven species of distributed from the western Mediterranean region to southwest Asia; and the sister genus *Catharanthus*, seven of eight species of which are Madagascar endemics.

The name "vinca," bestowed by Linnaeus in 1767, is derived from the name given in antiquity by Pliny the Elder, *vincaperivinca*, still reflected in its common Italian and French names, *pervinca* and *pervenche*, respectively. Two of the European *Vinca* species, *V. major* and *V. minor*, have long been prized horticulturally; both are lustrous-green creeping evergreen vines sporting lavender-colored pinwheel-shaped flowers. Because the plants are low growing and spread quickly, they are often used as groundcover in garden landscapes, but have become invasive in many parts of the world.

Darwin was curious about pollination and seed production in *Vinca*. He noticed that *Vinca major* never seemed to set seed in England and figured it was for want of the right insect pollinators. He did an experiment using a slender bristle to mimic a moth pollinator, described in a letter to Daniel Oliver, Keeper of the Herbarium at the Royal Botanic Gardens, Kew: "I have passed *fine* bristle between anthers (not cutting or touching the flowers) in same way as proboscis of moth would pass to nectary, near the sides of the corolla; pollen sticks to bristle and a bristle thus covered from pollen of *one* flower is used for another flower.

... I have 4 or 5 fine pods swelling."[150] He published the results in the *Gardeners' Chronicle*, concluding with his usual crowd-sourcing appeal, urging readers to try the experiment and send in the results.[151] At Darwin's urging, Charles Crocker, a Kew propagator always willing to help, experimented with the related tropical periwinkles. His results, equally as positive as Darwin's, were reported in the same journal the following month:

Following the suggestion made by Mr. Darwin at page 552, a week or two ago, I thought that I would try if the tropical kinds of Vinca could be induced to produce seed, which is never the case under cultivation if left to themselves. I impregnated eight flowers, and in the course of a few days had the satisfaction of seeing that the pistils in seven cases were swelling well. The erect double follicles are now in several instances more than an inch long; in one they are not yet ripe. The plant upon which I tried the experiment was the white-flowered variety of Vinca rosea {now Catharanthus roseus}. I used the pollen from the same plant as I wished also to see if this variety would reproduce itself by seed, or if it will revert to the normal colour of the species. I merely passed a hair down the tube of one flower after another as an insect might insert its proboscis in its search for nectar.[152]

Darwin's hunch was correct—outside of their native range, these ground-hugging plants have a hard time attracting pollinators with probosces long enough to fertilize them. Lack of seed production may be why these plants are not even more invasive, spreading vegetatively for the most part, and slowly at that.

> From: "Fertilisation of Vincas." *Gardeners' Chronicle*
> *and Agricultural Gazette* (15 June 1861, p. 552)

Fertilisation of Vincas.—I do not know whether any exotic *Vincas* seed, or whether gardeners would wish them to seed, and so raise new varieties. Having never observed the large Periwinkle or *Vinca major* to produce seed, and having read that this never occurs in Germany, I was led to examine the flower. The pistil, as botanists know, is a curious object, consisting of a style, thickening upwards, with a horizontal wheel on the top; and this is surmounted by a beautiful brush of white filaments. The concave tire of the wheel is the stigmatic surface, as was very evident when pollen was placed on it, by the penetration of the pollen-tubes. The pollen is soon shed out of the anthers and lies embedded in little alcoves in the white filamentous brush above the stigma. Hence it was clear that the pollen could not get on to the stigma without the aid of insects, which, as far as I have observed in England, never visit this flower. Accordingly, I took a fine bristle to represent the proboscis of a moth, and passed it down between the anthers, near the sides of the corolla; for I found that the pollen sticks to the bristle and is carried down to the viscid stigmatic surface. I took the additional precaution of passing it down first between the anthers of one flower and then of another, so as to give the flowers the advantage of a cross; and I passed it down between several of the anthers in each case. I thus acted on six flowers on two plants growing in pots; the germens of these swelled, and on four out of the six I have now got fine pods, above 1½ inch in length, with the seeds externally visible; whereas the flower stalks of the many other flowers all shanked off. I wish any one who wishes to obtain seed of any other species that does not habitually seed would try this simple little experiment and report the result. I shall sow the seeds of my *Vinca* for the chance of a sport: for a plant which seeds so rarely might be expected to give way to some freak on so unusual and happy an occasion.

Viola odorata. Watercolor by Elizabeth Wharton, *British Flowers.*

┌─────────────────────────────────┐
│   │
│   *Viola*   │
│   VIOLET   │
│   ───── ...... ─────   │
│   │
│   VIOLACEAE—VIOLET FAMILY   │
│   │
└─────────────────────────────────┘

FORMS OF FLOWERS, POLLINATION

Pansy, Johnny-jump-up and heartsease are common names for *Viola* varieties widely grown in gardens. In *Variation*, Darwin noted the history of *Viola* cultivation dating as far back as 1687, with many new varieties "energetically commenced" by nurserymen beginning in 1811. *Viola tricolor*, heartsease, was a Victorian-era staple of the flower garden, and the species from which most garden pansies were developed. It is therefore unsurprising, as a plant well familiar to his audience, that Darwin cited heartsease as an example of the tell-tale signs of artificial selection in *Origin*. The myriad cultivated heartsease varieties differ most in their flowers, the feature valued by people, he pointed out, but not much in their leaves:

> *There is another means of observing the accumulated effects of selection—namely, by comparing the diversity of flowers in the different varieties of the same species in the flower-garden; the diversity of leaves, pods, or tubers, or whatever part is valued, in the kitchen-garden ... See how different the leaves of the cabbage are, and how extremely alike the flowers; how unlike the flowers of the heartsease are, and how alike the leaves.*[153]

He delved in more deeply in *Variation*, tracing the steps by which garden-variety heartsease were developed: "The first great change was the conversion of the dark lines in the centre of the flower into a dark eye or centre, which at that period had never been seen, but is now considered one of the chief requisites of a first-rate flower."[154] Darwin went on to acknowledge the confusion of determining the parentage of cultivated varieties, with several descended from wild species, and more or less intercrossed to boot.

Darwin experimented with cross- and self-pollination in violets for several years beginning in the early 1860s, and he appreciated that the small, self-fertile, "cleistogamic" flowers found in many violet species assured abundant production of seeds. *Viola*, it happens, has perhaps the largest number of cleistogamous species of any flowering plant genus. Darwin dissected several of them, writing to Joseph Hooker in 1862, "I have been amusing myself by looking at the small flowers of *Viola* … What queer little flowers they are."[155]

[**From: *The Different Forms of Flowers on Plants of the Same Species* (1877)**]

It was known even before the time of Linnaeus that certain plants produced two kinds of flowers, ordinary open, and minute closed ones; and this fact formerly gave rise to warm controversies about the sexuality of plants. These closed flowers have been appropriately named cleistogamic by Dr. Kuhn. They are remarkable from their small size and from never opening, so that they resemble buds; their petals are rudimentary or quite aborted; their stamens are often reduced in number, with the anthers of very small size, containing few pollen-grains, which have remarkably thin transparent coats, and generally emit their tubes whilst still enclosed within the anther cells; and, lastly, the pistil is much reduced in size, with the stigma in some cases hardly at all developed. These flowers do not secrete nectar or emit any odour; from their small size, as well as from the corolla being rudimentary, they are singularly inconspicuous. Consequently, insects do not visit them; nor if they did, could they find an entrance. Such flowers are therefore invariably self-fertilised; yet they produce an abundance of seed.

In several cases, the young capsules bury themselves beneath the ground, and the seeds are there matured. These flowers are developed before, or after, or simultaneously with the perfect ones. Their development seems to be largely governed by the conditions to which the plants are exposed, for during certain seasons or in certain localities only cleistogamic or only perfect flowers are produced. ...

Viola canina.—The calyx of the cleistogamic flowers differs in no respect from that of the perfect ones. The petals are reduced to five minute scales; the lower one, which represents the lower lip, is considerably larger than the others, but with no trace of the spur-like nectary; its margins are smooth, whilst those of the other four scale-like petals are papillose. ... The stamens are very small, and only the two lower ones are provided with anthers, which do not cohere together as in the perfect flowers. The anthers are minute, with the two cells or loculi remarkably distinct; they contain very little pollen in comparison with those of the perfect flowers. The connective expands into a membranous hood-like shield which projects above the anther-cells. These two lower stamens have no vestige of the curious appendages which secrete nectar in the perfect flowers. The three other stamens are destitute of anthers and have broader filaments, with their terminal membranous expansions flatter or not so hood-like as those of the two antheriferous stamens. ... In the cleistogamic flowers, the pollen-grains, as far as I could see, never naturally fall out of the anther-cells, but emit their tubes through a pore at the upper end. I was able to trace the tubes from the grains some way down the stigma. The pistil is very short, with the style hooked, so that its extremity, which is a little enlarged or funnel-shaped and represents the stigma, is directed downwards, being covered by the two membranous expansions of the antheriferous stamens. It is remarkable that there is an open passage from the enlarged funnel-shaped extremity to within the ovarium; this was evident, as slight pressure caused a bubble of air, which had been drawn in by some accident, to travel freely from one end to the other: a similar passage was observed by Michalet in *V. alba*. The pistil therefore differs considerably from that of the perfect flower; for in the latter it is much longer, and straight with the exception of the rectangularly bent stigma; nor is it perforated by an open passage. ...

The seeds produced by the cleistogamic and perfect flowers do not differ in appearance or number. On two occasions, I fertilised several perfect flowers with pollen from other individuals and afterwards marked some cleistogamic flowers on the same plants; and the result was that 14 capsules produced by the perfect flowers contained on an average 9.85 seeds; and 17 capsules from the cleistogamic ones contained 9.64 seeds,—an amount of difference of no significance. It is remarkable how much more quickly the capsules from the cleistogamic flowers are developed than those from the perfect ones.

Vitis vinifera. Watercolor painting by Baldassare Cattrani,
in *Exoticarum atque indigenarum plantarum*.

```
┌─────────────────────────────────────────┐
│                                          │
│            Vitis vinifera                │
│            GRAPE VINE                     │
│          ──────  ······  ──────          │
│     VITACEAE—GRAPEVINE FAMILY            │
│                                          │
└─────────────────────────────────────────┘
```

Vitis vinifera

GRAPE VINE

VITACEAE—GRAPEVINE FAMILY

CLIMBING PLANTS

Vitaceae is a family of lianas—woody vines—with broad lobed leaves and tendrils and inflorescences that are opposite to the leaves, easy to recognize even in the absence of flowers or fruits. The word "vine" originally referred only to *Vitis* but now is considered a term for herbaceous climbing plants; "liana" refers to woody climbers. The genus *Vitis* includes around sixty species of grapes, twenty-five of which are found in North America.

The Old World species *Vitis vinifera* is well worth raising a glass to. Thought to have come into cultivation in the trans-Caucasus region by at least 6000 BCE, the locus of the oldest known viticultural tradition, it arrived in western Europe by around 1500 BCE and is now cultivated worldwide in regions with hot dry summers and cold wet winters. *Vitis* species and its hybrids are cultivated for the production of fresh table grapes, dried raisins, grape juice, and vinegar— even the leaves have their uses, such as stuffed dolma—but most significantly, wine production constitutes an estimated 80 percent of cultivation. Nowadays an estimated 10,000 cultivars are recognized, with many grafted onto the rootstocks of native North American grape species to protect them from the devastating grape phylloxera (*Daktulosphaira vitifoliae*), an aphid relative native to North America that almost wiped out the European vines.

Darwin was known to enjoy wine and port on occasion. In one record of undergraduate hijinks, he lost a bet on the height of the ceiling in the old Combination Room at Christ's College, Cambridge, costing him—or rather his father, who was footing his university bills—a bottle of port.* Later in life he got more serious, including on the subject of *Vitis*. He grew grapevines in his garden and studied the development of their tendrils closely. Documenting a graduated series from peduncle to tendril, Darwin came to believe that tendrils represent modified inflorescence stems, or peduncles—a departure from the tendrils of most species, which are mostly derived from leaves or stems. He also observed the related genera *Cissus* and *Ampelopsis* (now *Parthenocissus*), along with species in two other families, Sapindaceae and Passifloraceae, all of which, he concluded, also have tendrils derived from flower-peduncles.

This is accepted today, but Darwin had a difficult time convincing his botanist friends. Each had a different opinion on the origin of tendrils, at least certain kinds. Take gourds and others in the Cucurbitaceae family. Asa Gray was convinced their tendrils are modified branches; Darwin's old mentor John Stevens Henslow at Cambridge maintained that they are modified stipules; and botanist Thomas Thomson was sure they represent modified leaves. Darwin spent many months studying tendril development in different groups, comparing them closely to trace out homologies. At one point he was thinking all

---

* Darwin's son Francis reported this bet in *Life and Letters of Charles Darwin* (F. Darwin 1887, 1: 279–280):

"A trifling record of my father's presence in Cambridge occurs in the book kept in Christ's College combination-room, where fines and bets were recorded, the earlier entries giving a curious impression of the after-dinner frame of mind of the fellows. The bets were not allowed to be made in money, but were, like the fines, paid in wine. The bet which my father made and lost is thus recorded:—

'Feb. 23, 1837.—Mr. Darwin v. Mr. Baines, that the combination-room measures from the ceiling to the floor more than (*x*) feet. 1 Bottle paid same day. N.B. Mr. Darwin may measure at any part of the room he pleases.'"

were derived from leaves, writing to Hooker half-jokingly, "Every thing [would] go very beautifully for me if botanists [would] let all tendrils be modified leaves."[156] *Vitis* and *Passiflora* vines soon changed his mind. He ran his analysis by Daniel Oliver, at Kew, writing, "Does not this render it highly probable that the tendril is a modified flower with its peduncle?" Like the others, Oliver did not fully agree, and the conversation continued.[157] Darwin stuck to his position, concluding in *Climbing Plants* that while all tendrils performed the same function, they were, depending on the plant, derived from different organs, including leaves, flower-peduncles, and possibly branches and stipules.[158] Botanists studying tendril evolution today largely agree, recognizing some seventeen different kinds of tendril.[159]

From: *The Movements and Habits of Climbing Plants* (2nd ed., 1875)

Vitis vinifera.—The tendril is thick and of great length; one from a vine growing out of doors and not vigorously, was 16 inches long. It consists of a peduncle (A), bearing two branches which diverge equally from it. One of the branches (B) has a scale at its base; it is always, as far as I have seen, longer than the other and often bifurcates. The branches when rubbed become curved and subsequently straighten themselves. After a tendril has clasped any object with its extremity, it contracts spirally; but this does not occur when no object has been seized. The tendrils move spontaneously from side to side; and on a very hot day, one made two elliptical revolutions, at an average rate of 2 hrs. 15 m. During these movements, a coloured line, painted along the convex surface, appeared after a time on one side, then on the concave side, then on the opposite side, and lastly again on the convex side. The two branches of the same tendril have independent movements. After a tendril has spontaneously revolved for a time, it bends from the light towards the dark: I do not state this on my own authority, but on that of Mohl and Dutrochet. Mohl says that in a vine planted against a wall, the tendrils point towards it, and in a vineyard generally more or less to the north.

The young internodes revolve spontaneously; but the movement is unusually slight. A shoot faced a window, and I traced its course on the glass during two perfectly calm and hot days. On one of these days, it described, in the course of ten hours, a spire, representing two and a half ellipses. I also placed a bell-glass over a young Muscat grape in the hot-house, and it made each day three or four very small oval revolutions; the shoot moving less than half an inch from side to side. Had it not made at least three revolutions whilst the sky was uniformly overcast, I should have attributed this slight degree of movement to the varying action of the light. The extremity of the stem is more or less bent downwards, but it never reverses its curvature, as so generally occurs with twining plants.

Various authors believe that the tendrils of the vine are modified flower-peduncles. I here give a drawing (right) of the ordinary state of a young flower-stalk: it consists of the "common peduncle" (A); of the

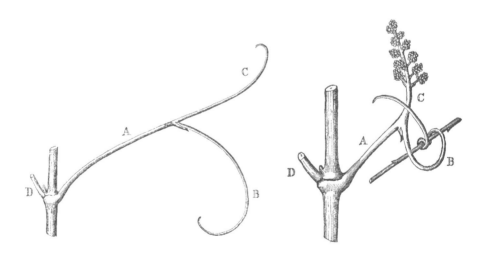

Tendril of the vine (left). Flower-stalk of the vine (right).

"flower-tendril" (B), which is represented as having caught a twig; and of the "sub-peduncle" (C) bearing the flower-buds. The whole moves spontaneously, like a true tendril, but in a less degree; the movement, however, is greater when the sub-peduncle (C) does not bear many flower-buds. The common peduncle (A) has not the power of clasping a support, nor has the corresponding part of a true tendril. The flower-tendril (B) is always longer than the sub-peduncle (C) and has a scale at its base; it sometimes bifurcates, and therefore corresponds in every detail with the longer scale-bearing branch (B, left) of the true tendril. It is, however, inclined backwards from the sub-peduncle (C), or stands at right angles with it, and is thus adapted to aid in carrying the future bunch of grapes. When rubbed, it curves and subsequently straightens itself; and it can, as is shown in the drawing, securely clasp a support. I have seen an object as soft as a young vine-leaf caught by one.

The lower and naked part of the sub-peduncle (C) is likewise slightly sensitive to a rub, and I have seen it bent round a stick and even partly round a leaf with which it had come into contact. That the sub-peduncle has the same nature as the corresponding branch of an ordinary tendril, is well shown when it bears only a few flowers; for in this case, it becomes less branched, increases in length, and gains both in sensitiveness and in the power of spontaneous movement. I have twice seen sub-peduncles which bore from thirty to forty flower-buds, and which had become considerably elongated and were completely wound round sticks, exactly like true tendrils. ...

The gradations from the ordinary state of a flower-stalk, as represented in the drawing (right), to that of a true tendril (left) are complete. We have seen that the sub-peduncle (C), whilst still bearing from thirty to forty flower-buds, sometimes becomes a little elongated and partially assumes all the characters of the corresponding branch of a true tendril. From this state, we can trace every stage till we come to a full-sized perfect tendril, bearing on the branch which corresponds with the sub-peduncle one single flower-bud! Hence there can be no doubt that the tendril is a modified flower-peduncle.

Oxalis versicolor. Watercolor on vellum by Pancrace Bessa for
Herbier Général de l'Amateur.

ON THE LIBRARY of
RACHEL LAMBERT MELLON
at OAK SPRING GARDEN
FOUNDATION

Throughout time, the interests of gardeners and botanical artists have been intimately intertwined. Both come together in the life of Rachel Lambert Mellon (1910–2014), the benefactor of Oak Spring Garden Foundation. As a young child, Rachel Lambert, nicknamed Bunny at birth, grew wildflowers in clay pots next to her bedroom window, and she developed garden designs based on illustrations from fairytale books. With the support of her father, Gerard B. Lambert (1886–1967), she created and planted a garden at Albemarle, the family home in Princeton, New Jersey. At age ten, she also began acquiring rare gardening books. This was the beginning of a collection she continued to build throughout her long life. Her early acquisitions, along with many other magnificent works collected over a period of more than nine decades, are the core of the library that Mrs. Mellon bequeathed to Oak Spring Garden Foundation.

This was a working library. Certainly, Mrs. Mellon was a collector, but she was also a reader and used her library in her research, her gardening practice, and her garden designs. Her first professional design commission was in 1933 for fashion designer Hattie Carnegie (1886–1956) in New York City. Other commissions followed for family and friends, but most significant was the request of President John F. Kennedy in 1961 for Mrs. Mellon to help design and plant the White House Rose Garden. Later design commissions included the grounds of the John F. Kennedy Presidential Library and the National Gallery of Art.

Her interest in horticulture and botany led Mrs. Mellon to collect beautiful paintings and drawings relating to plants and natural history, including works by old master, impressionist, post-impressionist, and modern painters. She acquired exquisite plant portraits by Rory McEwen, Margaret Stones, Sophie Grandval-Justice and other renowned botanical artists of the twentieth century. Her passion for the natural world also influenced her commissions of exceptional objects of art by well-known designers and sculptors, including Jean Schlumberger and Diego Giacometti, as well as textiles and clothes by fashion icons Cristóbal Balenciaga and Hubert de Givenchy.

Over the years, Mrs. Mellon had collected over 16,000 objects, from the fourteenth through the twentieth century, most relating to gardens, plants and natural history. Her rare and modern books, manuscripts, drawings, prints, and other decorative objects span from the fourteenth to the twentieth century. In 1976, she built Oak Spring Garden Library with the help of her husband, Paul Mellon (1907–1999), and in 1993, she established Oak Spring Garden Foundation.

Mrs. Mellon published four discursive catalogues summarizing different aspects of the Library's collection: *An Oak Spring Sylva* (1989), *An Oak Spring Pomona* (1990), *An Oak Spring Flora* (1997/1998) and *An Oak Spring Herbaria* (2009). These four volumes, which can be viewed on the Foundation's website, demonstrate Mrs. Mellon's commitment to sharing her extraordinary collection, which continues to attract gardeners, botanists, naturalists, authors, artists, and many others from all over the world. Mrs. Mellon would be delighted with this beautiful book, which connects a great scientist with great botanical art and makes creative use of the illustrations that she so enjoyed.

TONY WILLIS
Head Librarian
Oak Spring Garden Foundation

From the great variety of images contained in the rare books, manu-scripts, and drawings of the library of Oak Spring Garden Foundation, (OSGF) any of which could have been published alongside Darwin's text, we enjoyed selecting just a few of from the works listed below. Informal titles for unpublished manuscripts and illustrations are indicated by quotation marks. The names of plants from each work illustrated in this book are given in parentheses after each entry.

The Orchidaceae of Mexico and Guatemala—James Bateman (1811– 1897), one of the most renowned orchidologists of the nineteenth century, published this work in London between 1837 and 1843, during a time of European obsession with collecting and raising exotic orchids. The book is exquisite, luxurious and unusually large, 30 × 22 inches. Only 125 copies, each with forty hand-colored lithographs, were issued to subscribers. The illustrations, most newly discovered species, show the orchids life size and in full bloom. Publication continued over the six years as rare specimens came into bloom. The orchids were drawn by Sarah Anne Drake (1803–1857), Jane Edwards (1842–1898), Samuel Holden (active 1845–1847), and Augusta Innes Withers (1792–1877), and lithographed by Maxime Gauci (1774–1854). (*Catasetum maculatum*)

Phaseolus Brasilicus.

25

Phaseolus brasilicus. Hand-colored engraved plate by Johann Theodor de Bry in
Florilegium Renovatum et Auctum.

Herbier Général de l'Amateur—Pancrace Bessa (1772–1846) was one of the greatest flower painters of his era and was associated with the Muséum National d'Histoire Naturelle in Paris. He was trained and influenced by Pierre-Joseph Redouté and Gerard van Spaendonck, both of whom are also represented at Oak Spring. Bessa created 600 watercolors on vellum for eight volumes of this work by Jean Claude Michel Mordant de Launay published in Paris, 1816–1827, commissioned by King Charles X. The paintings were dispersed at auction in 1947, and OSGF has eighty-nine of the originals. (*Gloriosa superba, Oxalis versicolor*)

A Curious Herbal—Elizabeth Blackwell (1700–1758) took on the task of drawing medicinal plants at Chelsea Physic Garden in London to financially support her husband, Alexander, while he was in debtor's prison. Blackwell drew 500 species, many of which were plants discovered in the Americas, and she engraved the illustrations on copper. She wrote the descriptions and medicinal properties of each plant with her husband. *A Curious Herbal* was published with four plates weekly from 1737 to 1739, and she hand-colored some copies. Blackwell was one of the few eighteenth century women credited as both artist and publisher. Her work was successful and was soon re-engraved and republished in a larger two-volume version by Christoph Jacob Trew. Her paintings were purchased by the avid collector, John Stuart, Third Earl of Bute, a Prime Minister of Great Britain and a confidant of Princess Augusta whose botanic garden developed into the Royal Botanic Gardens, Kew. OSGF acquired seventy-two of Blackwell's original watercolors, along with two copies of the English edition of *A Curious Herbal* and the edition republished by Trew. (*Fragaria vesca, Orchis mascula, Pisum sativum*)

Florilegium Renovatum et Auctum—Johann Theodor de Bry (1561–1623), a German engraver and publisher, produced eighty-two hand-colored engravings for this work published in 1641 by his son-in-law Matthäus Merian, father of artist Maria Sibylla Merian. Many of de Bry's engravings were copied from the prints of other artisans, including that of *Phaseolus brasilicus*, which was drawn by Giovanni Battista Ferrari (1584–1655). (*Phaseolus brasilicus*)

Stirpium Imagines and *Exoticarum Atque Indigenarum Plantarum*—A contemporary of Pierre-Joseph Redouté, Baldassare Cattrani (fl. 1776–1810) was commissioned by Empress Josephine to create a collection of masterful body color paintings. Later Cattrani returned to his homeland to draw native and exotic flowering plants at the botanical garden in Padua, Italy, the oldest botanical garden in Europe. At OSGF, there are seventy-five original drawings bound in a manuscript entitled *Stirpium Imagines* (Padua, 1776), with sixty-three unbound drawings (not dated) that were drawn for *Exoticarum Atque Indigenarum Plantarum*, and a set of seven watercolors on vellum, circa 1799. (*Vitis vinifera*)

Flora Londinensis—William Curtis (1746–1799), with the title of Demonstrator of Botany to the Company of Apothecaries, left his apprenticeship as an apothecary and pharmacist to pursue his passion for plants. He labored over ten years on *Flora Londinensis*, the first work to comprehensively describe and depict plants that grew around London. Three volumes were published between 1777 and 1796, with 306 hand-colored engravings. The elegant originals were drawn by James Sowerby (1757–1822), Sydenham Teak Edwards (1768–1819), Francis Sansom (1780–1810), and William Kilburn (1745–1818). It was considered a masterpiece by his fellow naturalists but was not a financial success, and Curtis then realized that illustrations and texts on exotic plants would be more popular with the public, leading to his next publication, *The Botanical Magazine*. (*Linaria vulgaris, Primula acaulis*)

The Botanical Magazine—William Curtis (1746–1799) founded *The Botanical Magazine* in 1793 and successfully recouped his losses from *Flora Londinensis*. The first major journal on botany to be published in England, *The Botanical Magazine* has been published continuously and is now called *Curtis's Botanical Magazine*. From the very beginning, it featured ornamental plants, both native and introduced, that are cultivated in gardens and greenhouses. Among the editors of *The Botanical Magazine* from Darwin's era were directors of the Royal Botanic Gardens, Kew, Sir William Jackson Hooker and his son Sir Joseph Dalton Hooker. The volumes were first issued with hand-colored copperplate engravings, later succeeded by colored lithography and then by modern color printing. Most of the early original art was created by James Sowerby (1757–1822), Sydenham Edwards (1768–1819), Walter Hood Fitch (1817–1892) and William Jackson Hooker (1785–1865). (*Cardiospermum halicacabum, Maurandya scandens, Salvia coccinea, Spiranthes cernua*)

Mémoires pour Servir à l'Histoire des Plantes—Denis Dodart (1634–1707) was a French botanist and physician who did a great deal of botanical research, including studies on plant physiology and the influence of gravity on plants. In 1675, he wrote this work, which was illustrated with thirty-nine plates, most drawn directly from life and engraved by the French artist Nicolas Robert (1616–1685). Robert worked at the Jardin du Roi at Versailles and his engravings of flowers are considered among the finest of the early seventeenth century. (*Bignonia capreolata*)

Flowers, Moths, Butterflies and Shells and *Plantae et Papiliones Rariores*— Georg Dionysius Ehret (1708–1770) is considered one of the greatest botanical artists whose talent and knowledge united art and science. He worked for scientists and connoisseurs who commissioned numerous botanical paintings and engravings. Ehret traveled throughout Europe as a young man, working first as a gardener and draftsman, and his collaboration and lifelong friendship with the Swedish botanist

Carl von Linné (Linnaeus) helped build his reputation. In 1736, he settled in England and established himself as an artist and teacher. Oak Spring has this beautiful collection, *Flowers, Moths, Butterflies and Shells*, with thirty-seven water and body color paintings on vellum created between 1756 and 1769. Some of the paintings may have been intended for *Plantae et Papiliones Rariores*, which Ehret produced in London. His painting of *Pinguicula gesneri* (now considered a synonym of *P. vulgaris*) was most likely created for Linnaeus who coined the genus name based on the description of the leaves. (*Cyclamen europaeum, Dianthus caryophyllus, Pinguicula vulgaris*)

"The Venus Flytrap Manuscript"—John Ellis (1705–1776), an English linen merchant, botanist, zoologist, and author, was a correspondent and close friend of Linnaeus who described him as "a bright star of natural history." In 1769, Ellis sent Linnaeus a description of *Dionaea muscipula* along with his pen-and-ink illustration of a live fly trap that had been transported to London. Ellis's letter announced the formal introduction of carnivorous plants to Western scientific discourse. The published description, with a colored print of the Venus fly trap, was appended to the pamphlet printed in 1770 on the transoceanic shipment of plants entitled *Directions for bringing over Seeds and Plants from the East Indies and other distant Countries in a state of vegetation: Together with a catalogue of such foreign plants as are worthy of being encouraged in our American colonies, for the purposes of medicine, agriculture, and commerce. To which is added The Figure and Botanical Description of a new Sensitive Plant, called Dionaea muscipula, or Venus's Fly-trap.* The original drawing of the Venus fly trap, and Ellis' letter to Linnaeus describing the plant, is at OSGF. (*Dionaea muscipula*)

"Album of Garden Flowers, English School"—This eighteenth-century album of garden flowers attributed to the English School includes 151 water and body color paintings on vellum. The collection includes beautiful images of garden plants in full bloom that were grown in England through the seasons, spring to autumn. The highly skilled artist surely had some botanical training, as many of the paintings are inscribed with the Latin and common names. (*Lathyrus odoratus, Lupinus pilosus, Ophrys apifera, Oxalis versicolor, Solanum tuberosum, Tropaeolum majus*)

"An Album of Flowers, French School"—Created in the late eighteenth century, this French School album of flowers consists of fifty-five watercolors that were probably part of a gardener's or florist's catalogue. (*Vinca minor*)

"162 Drawings of Plants"—Dame Ann Hamilton created beautifully composed and wonderfully accurate drawings of plants in water and body color on vellum between 1752 and 1766. A daughter of a British Member of Parliament, she is an elusive artist and very little is known about her life. Her paintings are exceptional, and her style is suggestive of Ehret, who taught many young women of "good family" and may have instructed her on floral painting. The paintings are accompanied by inscriptions of the common names of the plants, and often with their scientific names. OSGF has 162 of Hamilton's paintings. (*Clematis repens, Ipomoea purpurea, Linum perenne, Mimosa pudica, Phaseolus coccineus, Primula veris*)

A Catalogue of English Plants Drawn after Nature—Lady Frances Howard executed the watercolors on vellum that are bound into this manuscript between 1762 and 1766. Little is known about Howard's life and work, but she was a beloved pupil of Ehret and had similar skills, both exceptional and memorable. She clearly spent many hours on her work, applying sharp details and vibrant colors to her floral portraits. She indexed common and Latin names of the plants that she illustrated, suggesting she was interested in the scientific and geographic aspects of the plants as well as the artistic endeavor. The album at Oak Spring includes ninety-four watercolors on vellum and four experimental works on blue paper. (*Digitalis purpurea, Drosera rotundifolia*)

Lindenia: Iconography of Orchids—This work was published in thirteen volumes between 1891 and 1897 by Belgian botanist Jean Jules Linden (1817–1898), with 304 chromolithographs. The illustrations were drawn by A. Goessens and printed by Peter de Pannemaeker. Linden traveled to Brazil and other countries in the Neotropics collecting bromeliads and other plants, especially orchids, for European horticulture. He made detailed studies of the conditions in their native habitat and, once back in Belgium, created greenhouses with varied conditions for their cultivation. (*Catasetum saccatum*)

The Botanical Cabinet Consisting of Coloured Delineations of Plants from All Countries with a Short Account of Each—Conrad Loddiges (1821–1865) published this work in twenty parts between 1817 and 1833 with engravings by George Cooke (1781–1834). The work covered plants from all over the world that were cultivated in the Loddiges' nursery at Hackney, London, and many of the illustrations were done by members of the Loddiges family. (*Mitchella repens*)

Recueil de plantes copiées d'après nature par de Saint Aubin dessinateur du Roy Louis XV—Charles Germain de Saint Aubin (1721–1786) produced this collection of more than 250 watercolors, body colors, pastels, and ink wash drawings over a period of almost fifty years, between 1736 and 1785. Saint Aubin had wide-ranging interests and constantly experimented with new styles and techniques. Most of the paintings are exquisite studies of flowering plants, and others are skillfully composed bouquets and trompe l'oeil paintings, as well as landscapes and depictions of butterflies, seashells, monkeys, and other subjects. (*Vicia faba*)

Reichenbachia: Orchids Illustrated and Described by F. Sander—Frederick
Sander (1847–1920) was born in Germany and migrated to England
in search of employment as a nurseryman. He saw his first orchid at
the age of twenty-one and then devoted his life's work to them. As a
grower of orchids for Queen Victoria, he introduced many new varieties and around two hundred species have been named in his honor.
Reichenbachia was a collaboration between Sander and English landscape painter Henry George Moon (1857–1905), who created most
of the illustrations over a period of about four years. The first volume
was published in 1888 and three more volumes were published at
two-year intervals. Moon carved woodcut blocks, and the plates were
printed as chromolithography and, in a few cases, colored by hand.
In addition to Moon, other illustrators who contributed to the project were W. H. Fitch, Alice H. Loch, George Hansen, Charles Storer,
J. Watton, and James Laid Macfarlane. (*Angraecum sesquipedale*)

"Botanical Manuscript with 265 Drawings of Plants"—Elizabeth
Pieth Schmitz, wife of German botanist Martin Pieth Schmitz,
worked in the latter part of the seventeenth century and created
265 water and body color paintings of plants. As inscribed on the
first folio, all were "drawn according to nature." The paintings are
mounted onto fifty-nine folios and were bound into the manuscript
around 1787, now housed at OSGF. (*Cyclamen europaeum*)

A Flora of the State of New York—John Torrey (1796–1873) collected,
described, and classified plant specimens throughout the United
States, and one of his protégés was Asa Gray, later at Harvard University and Charles Darwin's most important supporter in America.
Torrey produced *A Flora of the State of New York* as part of the Natural History of New York series he published in 1843—it has 165
lithographs in two volumes, which are hand-colored in many of the
published editions. In Torrey's preface, he acknowledged his artists,

writing, "Many of the earlier drawings were executed by Miss Agnes Mitchell; the remainder by Miss Elizabeth Pooley with the exception of a few that were done by Mr. Swinton. They were all respectable artists, but they were unaccustomed to making dissections of plants. The lithography was executed at the office of Mr. George Endicott." (*Echinocystis lobata*)

Plantae Rariores Quas Maximam Partem and *Plantae Selectae*—Christoph Jacob Trew (1695–1769) was a Nuremberg physician, bibliophile, and botanist who was both supportive and productive in working with botanical artists. He was a lifelong friend of Ehret, and had influenced him to study plants scientifically as well as artistically. *Plantae Rariores Quas Maximam Partem* was printed in Altdorf and Nuremberg between 1784 and 1795 as a supplement to *Plantae Selectae*, published by Trew between 1750 and 1773. Both publications are beautifully illustrated by several artists including Benedict Christian Vogel (1745–1825) and Magnus Melchian Payerlein (1716–1751), with one plate by Ehret. The engravings were done by Christopher Keller (1638–1707) and Adam Ludwig Wirsing (1734–1797). (*Arachis hypogaea*)

The Orchid Album—Robert Warner (1814–1896), working with his publisher and coauthor Benjamin Samuel Williams (1824–1890), produced *The Orchid Album* by subscription in eleven volumes between 1882 and 1897. Each volume contains forty-eight lithographs, a total of 528, all of which were created by John Nugent Fitch (1840–1927), a nephew of the prolific botanical artist Walter Hood Fitch. The lithographs were printed in color and also colored by hand, depicting many of the spectacular orchid species and varieties that were then the obsession of Europe and England. *The Gardeners' Chronicle*, where Darwin often published articles, described *The Orchid Album* as a "magnum opus ... which was projected with the idea of supplying a demand for illustrations of Orchidaceous plants, with botanical descriptions of the plants figured and notes on their cultivation. ... Its appearance was hailed with great satisfaction in horticultural circles throughout the world." (*Coryanthes maculata*)

British Flowers—Elizabeth and Margaret Wharton, two sisters, produced a manuscript of 320 watercolor and pencil drawings of native wildflowers, bound in two volumes. The drawings are each inscribed with dates ranging from 1793 to 1811, along with notes on the plants' locations. In a brief biography, Elizabeth is praised for her botanical skills, which are reflected in the detail of the flower drawings. It is uncertain how the work was divided between the two sisters, but it seems that some pieces were painted together, with Elizabeth creating the bulk of the art and Margaret contributing intermittently. They produced two other volumes, one on grasses, the other on seaweeds, in the period between 1792 and 1827. (*Cypripedium calceolus, Epipactis latifolia, Humulus lupulus, Pulmonaria officinalis, Trifolium pratense, Viola odorata*)

Herbarius Ad Virum Delineatus—Jan Withoos (1648–circa 1685) created this Dutch florilegium around 1670, which consists of around 263 flowers painted on vellum in three volumes. Johannes, son of the well-known painter Matthias Withoos (1627–1703), spent most of his life in the Netherlands and had a profound appreciation for the natural world. He created beautiful paintings of native flowers as well as cultivated species which are historically significant because they reflect the popularity of new plants brought to Holland from around the world. Withoos's paintings were acquired by the bibliophile Paulo van Uchelen, who had them bound in full vellum, most likely by the Amsterdam bookbinder Albert Magnus. (*Ipomoea, Passiflora caerulea*)

ACKNOWLEDGMENTS

The authors thank The New York Botanical Garden, in particular Susan Fraser, now retired Director of the LuEsther T. Mertz Library, not only for introducing us, but for presenting two exhibitions that influenced Bobbi greatly, planting the seeds that became this book: *Darwin's Garden: An Evolutionary Adventure* (2008) and *Redouté to Warhol: Bunny Mellon's Botanical Art* (2016). Our deepest appreciation, too, to Peter Crane, Oak Spring Garden Foundation President, Tony Willis, Head Librarian, and OSGF photographer Jim Morris for their keen support and enthusiasm, hosting Bobbi as she researched the botanical art selections (special thanks to Tony for being especially helpful in presenting the books and manuscripts for Bobbi to peruse), kindly providing high-quality images of the artwork, and contributing the fine Foreword and the sections on Bunny Mellon's collection and our botanical art selections.

We were in fine hands from start to finish with Timber Press, beginning with our editors, Tom Fischer and Andrew Beckman, for whose enthusiasm, support, and advice we are deeply grateful. Many thanks, too, to Project Editor Jacoba Lawson for her careful and thorough copyediting and help shepherding the book through production, and to Timber Press's designers for their fine efforts in crafting the look and feel of the book. We could not have asked for a better publishing team!

Last but not least, this project benefitted tremendously from the assistance and critical comments of numerous friends and colleagues. Many thanks to Gustavo Romero (Gray Herbarium of Harvard University) and Susan Fraser, Jacquelyn Kallunki, Robbin Moran, and Robert Naczi at the New York Botanical Garden for their invaluable comments on the manuscript, and to Samantha D'Acunto, Susksma Dittakavi, and Olga Marta at the Mertz Library for kindly providing scans of Darwin's woodcut images. Very special thanks, finally, to Leslie Costa and Matt Candeias (*In Defense of Plants*), who read the entire manuscript and provided numerous helpful comments and corrections—needless to say, any errors that may remain are our sole responsibility.

ENDNOTES

INTRODUCTION

1. Lankester 1896, 2: 4391–4392.

2. A. Gray to C. R. Darwin, 1 September 1863, DCP-LETT-4288; *Correspondence* 11: 614.

3. Litchfield 1915, 1: 51.

4. J. M. Rodwell to F. Darwin, 8 July 1882 (DAR 112: 94v); *Correspondence* 1: 125, n2.

5. C. R. Darwin to R. W. Darwin, 31 August 1831, DCP-LETT-110; *Correspondence* 1: 132.

6. C. R. Darwin, Galápagos Notebook, p. 30b; Chancellor and van Wyhe 2009, p. 418.

7. C. R. Darwin, 1845 (*Journal of Researches*), p. 392.

8. C. R. Darwin to J. D. Hooker, 12 December 1843, DCP-LETT-722; *Correspondence* 2: 419; J. D. Hooker to C. R. Darwin, [12 December 1843–11 January 1844], DCP-LETT-723; Correspondence 2: 421.

9. C. R. Darwin to J. D. Hooker, 5 June 1855, DCP-LETT-1693; *Correspondence* 5: 343.

10. Darwin 1862 [*Orchids*], pp. 1-2.

11. C. R. Darwin to J. D. Hooker, 19 June 1860, DCP-LETT-3290; *Correspondence* 13: 427.

12. C. R. Darwin to A. Gray, 23–24 July 1862, DCP-LETT-3662; *Correspondence* 10: 330.

13. L. Darwin 1929, p. 118.

14. M. S. Wedgwood to C. R. Darwin, [before 4 August 1862], DCP-LETT-3681; *Correspondence* 10: 351; C. R. Darwin to K. E. S. Wedgwood, L. C. Wedgwood, and M. S. Wedgwood, 4 August 1862, DCP-LETT-4373; *Correspondence* 10: 355. See also Wedgwood 1868.

15. C. R. Darwin to A. Gray, 28 October 1876, DCP-LETT-10656; *Correspondence* 24: 325.

16. Edens-Meir and Bernhardt 2014 for the authoritative treatment of Darwin's orchid research and orchid biology

17. Nora Barlow (ed.). 1958. The Autobiography of Charles Darwin 1809-1882. London: Collins, p. 134.

18. C. R. Darwin to J. D. Hooker, 14 July 1863, DCP-LETT-4241, Correspondence 11: 533; C. R. Darwin to D. Oliver, 18 July 1863, DCP-LETT-4244, Correspondence 11: 543; C. R. Darwin to A. Gray, 4 August 1863, DCP-LETT-4262, Correspondence 11: 581

19. Isnard and Silk 2009 give a concise overview of research on climbers from Darwin to the present day.

20. Darwin 1875 [Insectivorous Plants], p. 286

21. Darwin 1880 [*Movement*], pp. 572, 573.

CHAPTER 1—*ANGRAECUM*

22. C. R. Darwin to J. D. Hooker, 25 [and 26] January 1862, DCP-LETT-3411; *Correspondence* 10: 47–49.

23. Wallace 1867, p. 488. This paper includes an illustration of the predicted moth pollinating *A. sesquipedale*. See also Arditti et al. 2012 and Kritsky 1991 for interesting treatments of the comet orchid pollinator story.

CHAPTER 2—*ARACHIS*

24. C. R. Darwin to J. D. Hooker, 25 March 1878, DCP-LETT-11443; *Correspondence* 26: 138–139.

CHAPTER 3—*BIGNONIA*

25. Darwin 1875 [*Climbing Plants*] pp. 98–99.

26. C. R. Darwin to A. Gray, 28 May 1864, DCP-LETT-4511; *Correspondence* 12: 211–214.

CHAPTER 4—*CARDIOSPERMUM*

27. For accounts of Soapberry bug biology and research, see entomologist Scott Carroll's informative website on *Soapberry Bugs of the World* (soapberrybug.org) and Caroll and Loye 2012.

28. C. R. Darwin to J. D. Hooker, 31 May 1864, DCP-LETT-4516; *Correspondence* 12: 224–225.

CHAPTER 5—*CATASETUM*

29. See Romero-Gonzalez 2018 for an in-depth account of Darwin's interest in these orchids.

30. Hooker 1825, p. 91.

31. C. R. Darwin to J. D. Hooker, 11 October 1861, DCP-LETT-328; *Correspondence* 9: 301.

32. Darwin 1877 [*Orchids*] p. 178.

33. Schomburgk 1837, Darwin 1862a

34. Crüger 1864, p. 127.

35. C. R. Darwin to W. E. Darwin, 12 October 1861, DCP-LETT-3284; *Correspondence* 9: 302–304.

CHAPTER 6—*CLEMATIS*

36. C. R. Darwin to J. D. Hooker, 8 February 1864, DCP-LETT-4403; *Correspondence* 12: 41–42.

37. C. R. Darwin to J. D. Hooker, 5 April 1864, DCP-LETT-4450; *Correspondence* 12: 119–121.

CHAPTER 7—*COBAEA SCANDENS*

38. J. D. Hooker to C. R. Darwin, 21 July 1863, DCP-LETT-4225; *Correspondence* 11: 554–555.

39. DAR 157.2 (*Climbing Plants*, Experimental notes, 1863–1864), p. 5.

CHAPTER 8—*CORYANTHES*

40. See Ramírez 2009.

41. C. R. Darwin to A. Gray, 25 February 1864, DCP-LETT-4415; *Correspondence* 12: 60–61.

42. H. Crüger to C. R. Darwin, 21 January 1864, DCP-LETT-4394; *Correspondence* 12: 23–27; Crüger 1864.

43. Darwin 1866 [*Origin*] p. 229.

CHAPTER 10—*CYPRIPEDIUM*

44. C. R. Darwin to A. Gray, 5 June 1861, DCP-LETT-3176; *Correspondence* 9: 162–164.

45. A. Gray to C. R. Darwin, 9 December 1862, DCP-LETT-3850; *Correspondence* 10: 591–592.

46. C. R. Darwin to R. Trimen, 27 August 1863, DCP-LETT-4279; *Correspondence* 11: 605–607.

47. C. R. Darwin to A. Rawson, 2 April 1863, DCP-LETT-4072F; *Correspondence* 24: 458–459.

48. C. R. Darwin to A. Rawson, 6 June 1863, DCP-LETT-5563; *Correspondence* 18: 395.

CHAPTER 11—*DIANTHUS*

49. Focke 1913, Leapman 2001.

50. E. Darwin, *The Loves of the Plants*, notes to "Canto IV" (see Darwin 1806, p. 217).

51. Darwin 1876 [*Cross and Self Fertilisation*] p. 9.

CHAPTER 12—*DIGITALIS*

52. Littler 2019.

53. See Costa 2017, chapter 6 for an account of this and other experiments in cross-pollination by Darwin.

CHAPTER 13—*DIONAEA*

54. T. Slaughter, ed., *Bartram: Travels and Other Writings* (New York: Library of America, 1996), p. 17. The quotation is found on pp. xx–xxi in the original 1791 edition of Bartram's *Travels*. An electronic transcription of the 1791 text can be found in the Documenting the American South collection of the University of North Carolina-Chapel Hill Libraries (docsouth.unc.edu).

55. C. R. Darwin to D. Oliver, 29 September 1860, DCP-LETT-2941; *Correspondence* 8: 398.

56. *Insectivorous Plants (1875)*, p. 286. See chapter 8 in *Darwin's Backyard* (Costa 2017) for more on Darwin's adventures with Venus's fly-trap and other carnivorous plants, and Forterre et al. 2005 for an innovative study of trap closure using high-speed video and microscopy.

CHAPTER 14—*DROSERA*

57. C. R. Darwin to J. D. Hooker, 11 September 1862, DCP-LETT-3721; *Correspondence* 10: 401–404.

58. Litchfield 1915, 2: 177.

59. C. R. Darwin to A. Gray, 26 September 1860, DCP-LETT-2930; *Correspondence* 8: 388–391.

60. C. R. Darwin to D. Oliver, 15 September 1860, DCP-LETT-2917; *Correspondence* 8: 357–358.

61. C. R. Darwin to A. Gray, 4 August 1863, DCP-LETT-4262; *Correspondence* 11: 581–584.

CHAPTER 15—*ECHINOCYSTIS*

62. Gray 1858, p. 98.

63. A. Gray to C. R. Darwin, 24 November 1862, DCP-LETT-3823; *Correspondence* 10: 553–555.

64. C. R. Darwin to J. D. Hooker, 25 June 1863, DCP-LETT-4221; *Correspondence* 11: 506–507.

65. C. R. Darwin to J. D. Hooker, 25 June 1863, DCP-LETT-4221; *Correspondence* 11: 506–507.

66. Darwin 1862 [*Climbing Plants*] p. 133

CHAPTER 16—*EPIPACTIS*

67. C. R. Darwin to A. G. More, 9 August 1860, DCP-LETT-2894; *Correspondence* 8: 316–317.

68. Darwin 1862 [*Orchids*] p. 39.

69. Darwin 1863a (van Wyhe 2009, p. 338).

70. Darwin 1877 [*Orchids*] pp. 101–102.

71. Jakubska et al. 2005.

CHAPTER 17—*FRAGARIA*

72. Darwin 1862b (van Wyhe 2009, pp. 322–323); see also *Correspondence* 10: 559.

73. For more on Darwin's experimental technique tracing movement see David Kohn's article in *Darwin's Garden: An Evolutionary Adventure* (Kohn 2008), and Mea Allen's *Darwin and His Flowers* (Allen 1977), p. 279.

CHAPTER 18—*GLORIOSA*

74. See Janet Browne's "Botany for gentlemen" (1989) for an interesting analysis.

75. Darwin 1877 [*Forms of Flowers*] pp. 146–147.

76. C. R. Darwin to J. D. Hooker, 14 July 1863, DCP-LETT 4241; *Correspondence* 11: 533–535.

77. DAR 157.1: 121.

78. Darwin 1882 [*Climbing Plants*] p. 194.

CHAPTER 19—*HUMULUS*

79. See Almaguer et al. 2014 for an overview of brewing with hops.

CHAPTER 20—*IPOMOEA*

80. See *The Effects of Cross and Self Fertilisation in the Vegetable Kingdom* (1878), pp. 15–18; C. R. Darwin to F. Galton, 13 January 1876, DCP-LETT-10357; *Correspondence* 24: 14–16.

81. Darwin 1878 [*Cross and Self Fertilisation*] p. 439.

CHAPTER 21—*LATHYRUS*

82. F. Darwin to C. R. Darwin, 14 August 1873, DCP-LETT-9009F; *Correspondence* 21: 329–330; C. R. Darwin to F. Darwin, 15 August 1873, DCP-LETT-9014; *Correspondence* 21: 334–336; F. Darwin to C. R. Darwin, 25 August 1873, DCP-LETT-9016; *Correspondence* 21: 348–349.

83. Hooker 1844, 1: 260–261.

84. C. R. Darwin to F. Delpino, 1 May 1873, DCP-LETT-8892; *Correspondence* 21: 198–199; F. Delpino to C. R. Darwin, 18 June 1873, DCP-LETT-8945; *Correspondence* 21: 258–260; C. R. Darwin to F. Delpino, 25 June 1873, DCP-LETT-8951; *Correspondence* 21: 264.

CHAPTER 22—*LINARIA*

85. Jachuła et al. 2018.

86. Darwin 1841 (van Wyhe 2009, pp. 134–137).

87. Leonard et al. 2013.

CHAPTER 23—*LINUM*

88. Darwin 1863b.

89. Darwin 1877 [*Forms of Flowers*] p. 90.

CHAPTER 24—*LUPINUS*

90. Darwin's lupine pollination observations are found in the "Torn-apart notebook," entries T103, 104, and 111; see Barrett et al. 1987, p. 457.

91. C. R. Darwin to J. D. Hooker, 22 August 1862, DCP-LETT-3696; *Correspondence* 10: 376–378.

92. Darwin 1880 [*Power of Movement*] p. 343.

CHAPTER 25—*MAURANDYA*

93. Darwin 1875 [*Climbing Plants*] p. 198.

94. A. Gray to C. R. Darwin, 28 December 1875, DCP-LETT-10329; *Correspondence* 23: 515–516; C. R. Darwin to A. Gray, 28 January 1876, DCP-LETT-10370; *Correspondence* 24: 31–32.

CHAPTER 26—*MIMOSA*

95. E. Darwin 1791 (*The Loves of the Plants*, "Canto I," lines 301–302).

96. C. R. Darwin to J. D. Hooker, 12 September 1873, DCP-LETT-9052; *Correspondence* 21: 378–379.

97. Darwin 1880 [*Power of Movement*] p. 395.

CHAPTER 27—*MITCHELLA*

98. Bell 1997, pp. 138-148.

99. A. Gray to C. R. Darwin, 11 October 1861, DCP-LETT-3282; *Correspondence* 9: 298–301.

100. C. R. Darwin to A. Gray, 2 January 1863, DCP-LETT-3897; *Correspondence* 11: 1–4.

101. Meehan 1868.

102. Darwin 1877 [*Forms of Flowers*] pp. 285, 287.

103. Hicks et al. 1985.

CHAPTER 28—*OPHRYS*

104. Ayasse 2009, Schiestl 2005.

105. C. R. Darwin to J. T. Moggridge, 13 October 1865, DCP-LETT-4914; *Correspondence* 13: 269–270.

106. C. R. Darwin to A. G. More, 17 July 1861, DCP-LETT-3211; *Correspondence* 9: 206–207.

CHAPTER 29—*ORCHIS*

107. Darwin 1877 [*Orchids*] p. 284.

108. C. R. Darwin to A. Gray, 23–24 July 1862, DCP-LETT-3662; *Correspondence* 10: 330–334.

109. C. R. Darwin to J. D. Hooker, 5 June 1860, DCP-LETT-2821; *Correspondence* 8: 237–240.

110. C. R. Darwin to A. G. More, 5 September 1860, DCP-LETT-2906; *Correspondence* 8: 343–345.

CHAPTER 30—*OXALIS*

111. G. Bentham to C. R. Darwin, 29 November 1861, DCP-LETT-3332; *Correspondence* 9: 353–354.

112. C. R. Darwin to J. D. Hooker, 25 March 1878, DCP-LETT-11443; *Correspondence* 26: 138–139.

CHAPTER 31—*PASSIFLORA*

113. C. R. Darwin to J. D. Hooker, 14 July 1863, DCP-LETT 4241; *Correspondence* 11: 533–535.

114. C. R. Darwin to D. Oliver, 11 March 1864, DCP-LETT-4424; *Correspondence* 12: 68–70.

115. C. R. Darwin to D. Oliver, 4 May 1864, DCP-LETT-4481; *Correspondence* 12: 165–166.

116. T. H. Farrer to C. R. Darwin, 29 June 1870, DCP-LETT-7254; *Correspondence* 18: 188–189.

CHAPTER 32—*PHASEOLUS*

117. *Darwin* 1841 (van Wyhe 2009, p. 136).

118. *Darwin* 1857, 1858 (van Wyhe 2009, pp. 267–268, 272–277).

119. *Darwin* 1858 (van Wyhe 2009, pp. 272–277).

CHAPTER 33—*PINGUICULA*

120. C. R. Darwin to A. Gray, 3 June 1874, DCP-LETT-9480; *Correspondence* 22: 275–276.

121. C. R. Darwin to W. T. Thiselton-Dyer, 23 June 1874, DCP-LETT-9508; *Correspondence* 22: 308–309.

122. Darwin 1888 [*Insectivorous Plants*] p. 315.

CHAPTER 34—*PISUM*

123. Bateson and Janeway 2008, pp. 94–95.

124. See "Questions for William Herbert," 1 April 1839, DCP-LETT-502; *Correspondence* 2: 179–182; and Herbert's reply: W. Herbert to J. S. Henslow, 5 April 1839, DCP-LETT-503; *Correspondence* 2: 182–185.

125. See Barrett et al. 1987, Q&E 11 and T151 (pp. 501, 469). See also Abberley & R. Darwin to C. R. Darwin, 18 October 1841, DCP-LETT-610; *Correspondence* 2: 306.

CHAPTER 35—*PRIMULA*

126. *Origin*, pp. 49–50; see also Costa 2009 for additional explanatory notes on this passage.

127. Huxley 1870, p. 402.

128. *Autobiography* (Barlow 1958), p. 134.

CHAPTER 36—*PULMONARIA*

129. Darwin 1862c, 1863b

130. C. R. Darwin to A. Gray, 19 Oct 1865, DCP-LETT-4919; *Correspondence* 13: 274–277.

CHAPTER 37—*SALVIA*

131. Darwin 1876 [*Cross and Self Fertilisation*] p. 5

132. F. Hildebrand to C. R. Darwin, 21 June 1864, DCP-LETT-4542; *Correspondence* 12: 254–255; C. R. Darwin to F. Hildebrand, 25 June 1864, DCP-LETT-4545; *Correspondence* 12: 256–257. Hildebrand published the results of this research in volume 4 of the *Jahrbücher für wissenschaftliche Botanik* (Hildebrand 1866).

CHAPTER 38—*SOLANUM*

133. Darwin 1845, p. 285.

134. See *Global Plants* biography of Carlos Ochoa (plants.jstor.org/stable/10.5555/al.ap.person.bm000025683).

135. Darwin 1875 [*Variation*] 1: 350.

136. Bateson and Janeway 2008, pp. 96–97.

CHAPTER 39—*SPIRANTHES*

137. C. R. Darwin to A. Gray, 31 October 1860, DCP-LETT-2969; *Correspondence* 8: 451–454; A. Gray to C. R. Darwin, 5 September 1862, DCP-LETT-3712; *Correspondence* 10: 394–395.

CHAPTER 40—*TRIFOLIUM*

138. C. R. Darwin to J. D. Hooker, 11 September 1859, DCP-LETT-2490; *Correspondence* 7: 332. The results of this experiment are also given in Darwin's Experimental notebook (DAR 157a), pp. 46–47.

139. Carreck et al. 2009.

140. For additional explanatory notes on these passages, see Costa 2009.

141. C. Hardy to C. R. Darwin, 23 July 1860, DCP-LETT-2877; *Correspondence* 8: 300–301; C. R. Darwin to C. Hardy, 27 July 1860, DCP-LETT-2879; *Correspondence* 8: 301.

142. C. R. Darwin to J. Lubbock, 2 September 1862, DCP-LETT-3708; *Correspondence* 10: 387–388 and 3 September 1862, DCP-LETT-3705; *Correspondence* 10: 392.

143. C. R. Darwin to F. Darwin, 25 July 1878, DCP-LETT- 11631; *Correspondence* 26: 320–322.

CHAPTER 41—*TROPAEOLUM*

144. Elisabet Christina Linnaea's paper "Om Indianska Krassens Blickande" ("On the twinkling of Indian Cress") was published in volume 23 of *Kongl. Vetenskaps Academiens Handlingar* (*Acts of the Royal Swedish Academy of Sciences*), 1763, 23: 284–286, along with commentary by physicist Johan Carl Wilcke on electrical phenomena (1763, 23: 286–287).

145. E. Darwin: *The Loves of the Plants*, "Canto IV," lines 43–51 (see E. Darwin 1806, pp. 191–192).

146. S. T. Coleridge, "Lines Written At Shurton Bars, Near Bridgewater, September, 1795, In Answer To A Letter From Bristol," published in *Poems on Various Subjects* (Coleridge 1796).

147. See Blick 2017 and references therein.

CHAPTER 42—*VICIA FABA*

148. Darwin 1888 [*Power of Movement*] p. 573.

149. Brenner et al. 2006, Baluška et al. 2009.

CHAPTER 43—*VINCA*

150. C. R. Darwin to D. Oliver, 27 May 1861, DCP-LETT-3161; *Correspondence* 9: 145–146.

151. Darwin 1861 (van Wyhe 2009, pp. 311–312).

152. Crocker 1861, p. 699.

CHAPTER 44—*VIOLA*

153. Darwin 1859 [*Origin*] p. 33; see also Costa 2009.

154. Darwin 1868 [*Variation*] 1: 368.

155. C. R. Darwin to J. D. Hooker, 30 May 1862, DCP-LETT-3575; *Correspondence* 10: 226–227.

CHAPTER 45—*VITIS*

156. C. R. Darwin to J. D. Hooker, 27 January 1864, DCP-LETT-4398; *Correspondence* 12: 31–33.

157. C. R. Darwin to D. Oliver, 11 March 1864, DCP-LETT-4424; *Correspondence* 12: 68–70; D. Oliver to C. R. Darwin, 12 March 1864, DCP-LETT-4425; *Correspondence* 12: 71–73.

158. Darwin 1877 [*Climbing Plants}* pp. 110–111.

159. Sousa-Baena et al. 2018.

BIBLIOGRAPHY

DIGITAL RESOURCES

Darwin Correspondence Project (darwinproject.ac.uk): University of Cambridge–based searchable digital archive of the approximately 15,000 letters of Charles Darwin, identified by "DCP-LETT" numbers. The DCP site also includes biographies of correspondents, historical documents, and treatments of Darwin in relation to a range of topics such as family, religion, and research activities.

Darwin Online (darwin-online.org.uk): Comprehensive digital collection of Darwin's publications directed by John van Wyhe (National University of Singapore), including books (in all editions), articles, printed letters, and published manuscripts. This site also presents the largest collection of unpublished Darwin manuscripts and private papers, identified by their "DAR" numbers, most made available courtesy of Cambridge University Library.

Oak Spring Garden Foundation Library (osgf.org/library): Bunny Mellon's collection of over 19,000 objects, including rare books, manuscripts, and works of art dating back to the fourteenth century. The collection mainly encompasses works relating to horticulture, landscape design, botany, natural history and voyages of exploration. There are also components relating to architecture, decorative arts, and classical literature.

LITERATURE CITED

Allen, M. 1977. *Darwin and his Flowers: The Key to Natural Selection.* London: Faber and Faber.

Almaguer, C., C. Schönberger, M. Gastl, E. K. Arendt, and T. Becker. 2014. "*Humulus lupulus*—a story that begs to be told. A review." *Journal of the Institute of Brewing* 120(4): 289–314.

Arditti, J., J. Elliott, I. J. Kitching, and L. T. Wasserthal. 2012. "'Good Heavens what insect can suck it'— Charles Darwin, *Angraecum sesquipedale* and *Xanthopan morganii praedicta.*" *Botanical Journal of the Linnean Society* 169: 403–432.

Ayasse, M. 2009. "Chemical mimicry in sexually deceptive orchids of the genus *Ophrys.*" *Phyton* 46(2): 221–223.

Baluška, F., S. Mancuso, D. Volkmann, and P. W. Barlow. 2009. "The 'root-brain' hypothesis of Charles and Francis Darwin: Revival after more than 125 years." *Plant Signaling & Behavior* 4(12): 1121–1127.

Barlow, N. (ed.). 1958. *The Autobiography of Charles Darwin 1809–1882. With the Original Omissions Restored. Edited and with Appendix and Notes by his Grand-daughter Nora Barlow.* London: Collins.

Barrett, P. H., P. J. Gautrey, S. Herbert, D. Kohn, and S. Smith (eds.). 1987. *Charles Darwin's Notebooks, 1836–1844.* Ithaca, NY: Cornell University Press.

Bateman, J. 1837–1843. *Orchidaceae of Mexico and Guatemala.* London: J. Ridgway & Sons.

Bateson, D. and W. Janeway. 2008. *Mrs. Charles Darwin's Recipe Book*. New York: Glitterati Inc.

Bell, W. J. 1997. "Patriot-Improvers: Biographical Sketches of Members of the American Philosophical Society. Volume 1. 1743–1768." *Memoirs of the American Philosophical Society* v. 226. Philadelphia: American Philosophical Society.

Blick, F. 2017. "Flashing flowers and Wordsworth's 'Daffodils.'" *Wordsworth Circle* 48(2): 110–115.

Brenner, E. D., R. Stahlberg, S. Mancuso, J. Vivanco, F. Baluška, and E. Van Volkenburgh. 2006. "Plant neurobiology: An integrated view of plant signaling." *Trends in Plant Science* 11(8): 413–419.

Browne, J. 1989. "Botany for gentlemen: Erasmus Darwin and 'The Loves of the Plants.'" *Isis* 80(4): 593–621.

Burkhardt, F., J. Secord, et al. (eds.), 1985. *The Correspondence of Charles Darwin*. 30 vols. Cambridge, UK: Cambridge University Press.

Carreck, N., T. Beasley, and R. Keynes. 2009. "Charles Darwin, cats, mice, bumblebees and clover." *Bee Craft*. (February 2009): 4–6.

Carroll, S. P. and A. E. Loye. 2012. "Soapberry bug (Hemiptera: Rhopalidae: Serinethinae) native and introduced host plants: Biogeographic background of anthropogenic evolution." *Annals of the Entomological Society of America* 105(5): 671–684.

Chancellor, G. and J. van Wyhe (eds.). 2009. *Charles Darwin's Notebooks from the Voyage of the Beagle*. Cambridge, UK: Cambridge University Press.

Coleridge, S. T. 1796. "Lines Written At Shurton Bars, Near Bridgewater, September, 1795 In Answer To A Letter From Bristol." *Poems on Various Subjects*. London: G. G. and J. Robinson, and Bristol: J. Cottle.

Correspondence. *The Correspondence of Charles Darwin* (F. Burkhardt et al., eds.) Volumes 1–30. Cambridge, UK: Cambridge University Press.

Costa, J. T. 2009. *The Annotated Origin: A Facsimile of the First Edition of On the Origin of Species*. Cambridge, MA: Belknap Press of Harvard University Press

Costa, J. T. 2014. *Wallace, Darwin, and the Origin of Species*. Cambridge, MA: Harvard University Press.

Costa, J. T. 2017. *Darwin's Backyard: How Small Experiments Led to a Big Theory*. New York: W. W. Norton.

Crocker, C. W. 1861. "Fertilisation of *Vinca rosea*." *Gardeners' Chronicle and Agricultural Gazette*, 27 July: p. 699.

Crüger, H. 1864. "A few notes on the fecundation of orchids and their morphology." Read 3 March. *Journal of the Proceedings of the Linnean Society of London*, Botany, 8: 127–135, pl. 9.

Darwin, C. R. 1841. "Humble-bees." *Gardeners' Chronicle and Agricultural Gazette*, no. 34; 21 August: 550.

Darwin, C. R. 1845. *Journal of Researches into the Natural History and Geology of the Countries Visited During the Voyage of H.M.S. Beagle Round the World* (2nd ed.). London: John Murray.

Darwin, C. R. 1857. "Bees and the fertilisation of kidney beans." *Gardeners' Chronicle and Agricultural Gazette* no. 43; 24 October: 725.

Darwin, C. R. 1858. "On the agency of bees in the fertilisation of papilionaceous flowers, and on the crossing of kidney beans." *Gardeners' Chronicle and Agricultural Gazette* no. 46; 13 November: 828–829.

Darwin, C. R. 1859. *On the Origin of Species by Means of Natural Selection, or the Preservation of Favoured Races in the Struggle for Life*. 1st ed. London: John Murray.

Darwin, C. R. 1860. "Fertilisation of British orchids by insect agency." *Gardeners' Chronicle and Agricultural Gazette* no. 23; 9 June: 528.

Darwin, C. R. 1861. "Fertilisation of Vincas." *Gardeners' Chronicle and Agricultural Gazette* no. 24; 15 June: 552.

Darwin, C. R. 1862a. *The Various Contrivances by which Orchids are Fertilised by Insects*. London: John Murray.

Darwin, C. R. 1862b. "On the three remarkable sexual forms of Catasetum tridentatum, an orchid in the possession of the Linnean Society." Read 3 April. *Journal of the Proceedings of the Linnean Society of London*, Botany, 6: 151–157.

Darwin, C. R. 1862c. "Cross-breeds of strawberries." (Letter). *Journal of Horticulture, Cottage Gardener, and Country Gentleman* 3 n.s.; 25 November: 672.

Darwin, C. R. 1862d. "On the two forms, or dimorphic condition, in the species of Primula, and on their remarkable sexual relations." Read 21 November. *Journal of the Proceedings of the Linnean Society of London*, Botany, 6: 77–96.

Darwin, C. R. 1863a. "Appearance of a plant in a singular place." *Gardeners' Chronicle and Agricultural Gazette* no. 33; 15 August: 773.

Darwin, C. R. 1863b. "On the existence of two forms, and on their reciprocal sexual relation, in several species of the genus Linum." *Journal of the Proceedings of the Linnean Society of London*, Botany, 7: 69–83.

Darwin, C. R. 1865. "On the movements and habits of climbing plants." *Journal of the Linnean Society*, Botany, 9: 1–118.

Darwin, C. R. 1866. *On the Origin of Species by Means of Natural Selection, or the Preservation of Favoured Races in the Struggle for Life*. 4th ed. London: John Murray.

Darwin, C. R. 1868. *The Variation of Animals and Plants Under Domestication*. 2 vols. 1st ed. London: John Murray.

Darwin, C. R. 1875a. *Insectivorous Plants*. 1st ed. London: John Murray.

Darwin, C. R. 1875b. *The Movements and Habits of Climbing Plants*. 2nd ed. London: John Murray.

Darwin, C. R. 1875c. *The Variation of Animals and Plants Under Domestication*. 2 vols. 2nd ed. London: John Murray.

Darwin, C. R. 1876. *The Effects of Cross and Self Fertilisation in the Vegetable Kingdom*. 1st ed. London: John Murray.

Darwin, C. R. 1877a. *The Different Forms of Flowers on Plants of the Same Species*. 1st ed. London: John Murray.

Darwin, C. R. 1877b. *The Various Contrivances by which Orchids are Fertilised by Insects*. 2nd ed. London: John Murray.

Darwin, C. R. 1878. *The Effects of Cross and Self Fertilisation in the Vegetable Kingdom*. 2nd ed. London: John Murray.

Darwin, C. R. 1880. *The Power of Movement in Plants*. 1st ed. London: John Murray.

Darwin, C. R. 1888. *Insectivorous Plants*. (Revised by Francis Darwin). 2nd ed. London: John Murray.

Darwin, E. 1806. *The Poetical Works of Erasmus Darwin* vol. 2: London: J. Johnson.

Darwin, F. 1880. "Experiments on the nutrition of *Drosera rotundifolia*." *Journal of the Linnean Society of London*, Botany, 17: 17–32.

Darwin, F. (ed.) 1887. *The Life and Letters of Charles Darwin, Including an Autobiographical Chapter* Volume 1. London: John Murray.

Darwin, L. 1929. "Memories of Down House." *The Nineteenth Century* 106: 118–123.

Edens-Meir, R. and P. Bernhardt (eds.). 2014. *Darwin's Orchids, Then and Now.* Chicago and London: University of Chicago Press.

Focke, W. O. 1913. "History of plant hybrids." *Monist* 23(3): 396–416.

Forterre, Y., J. M. Skotheim, J. Dumais, and L. Mahadevan. 2005. "How the Venus flytrap snaps." *Nature* 433: 421–425.

Gray, A. 1858. "Note on the coiling of tendrils." *Proceedings of the American Academy of Arts and Sciences* 4 (May 1857–May 1860): 98–99.

Hicks, D. J., R. Wyatt, and T. R. Meagher. 1985. "Reproductive biology of distylous Partridgeberry, *Mitchella repens.*" *American Journal of Botany* 72(10): 1503–1514.

Hildebrand, F. 1866. "Über die Befruchtung der Salviaarten mit Hülfe von Insekten." *Jahrbücher für wissenschaftliche Botanik*, Band 4 (1865–1866), 451–478.

Hooker, J. D. 1844. *The Botany of the Antarctic Voyage of H.M. Discovery Ships* Erebus *and* Terror *in the Years 1839–1843, Under the Command of Captain Sir James Clark Ross* Volume 1. London: Reeve Brothers.

Hooker, W. J. 1825. *Exotic Flora, Containing Figures and Descriptions of New, Rare, or Otherwise Interesting Exotic Plants.* Volume 2. Edinburgh: William Blackwood.

Huxley, T. H. 1870. (Address of Thomas Henry Huxley, LL.D., F.R.S., President) *Nature* 2: 400–406.

Isnard, S. and W. K. Silk. 2009. "Moving with climbing plants from Charles Darwin's time into the 21st century." *American Journal of Botany* 96(7): 1205–1221.

Jachuła, J., A. Konarska, and B. Denisow. 2018. "Micromorphological and histochemical attributes of flowers and floral reward in *Linaria vulgaris* (Plantaginaceae)." *Protoplasma* 255: 1763–1776.

Jakubska, A., D. Przado, M. Steininger, J. Aniol-Kwiatkowska, and M. Kadej. 2005. "Why do pollinators become 'sluggish'? Nectar chemical constituents from *Epipactis helleborine* (L.) Crantz (Orchidaceae)." *Applied Ecology and Environmental Research* 3(2): 29–38.

Kohn, D. 2008. *Darwin's Garden: An Evolutionary Adventure.* (Exhibition Catalog) New York Botanical Garden.

Kritsky, G. 1991. "Darwin's Madagascan hawk moth prediction." *American Entomologist* 37: 206–209.

Lankester, E. R. 1896. "Charles Robert Darwin." In *Library of the World's Best Literature Ancient and Modern*, edited by C. D. Warner. Volume 2: 4391–4392. New York: R. S. Peale & J. A. Hill.

Leapman, M. 2001. *The Ingenious Mr. Fairchild: The Forgotten Father of the Flower Garden.* New York: St. Martin's Press.

Leonard, A. S., J. Brent, D. R. Papaj, and A. Dornhaus. 2013. "Floral nectar guide patterns discourage nectar robbing by bumble bees." *PLoS One* 8: e55914.

Linnaea, E. C. 1763. "Om Indianska Krassens Blickande. ['On the twinkling of Indian Cress']" *Kongl. Vetenskaps Academiens Handlingar For Ar. 1762.* 23: 284–286.

Litchfield, H. E. (ed.). 1915. *Emma Darwin: A Century of Family Letters, 1792–1896* 2 Volumes. London: John Murray.

Littler, W. A. 2019. "Withering, Darwin and digitalis." *QJM: An International Journal of Medicine* 112: 887–890.

Meehan, T. 1868. *"Mitchella repens*, L., a dioecious plant." *Proceedings of the Academy of Natural Sciences of Philadelphia* 20: 183–184.

von Mohl, H. 1827. *Ueber den Bau und das Winden der Ranken und Schlingpflanzen.* Tübingen: Heinrich Laupp.

Müller, H. 1873. "Fertilisation of flowers by insects. III. On the co-existence of two forms of flowers in the same species or genus …" *Nature* 8: 433–435.

Müller, H. 1873b. "Fertilisation of flowers by insects. IV. On the two forms of flower of *Viola tricolor*, and on their different mode of fertilisation." *Nature* 9: 44–46.

Ramírez, S. R. 2009. "Orchid bees." *Current Biology* 19(23): R1062.

Romero-González, G. A. 2018. "Charles Darwin on Catasetinae (Cymbidieae, Orchidaceae)." *Harvard Papers in Botany* 23(2): 339–379.

Schiestl, F. P. 2005. "On the success of a swindle: Pollination by deception in orchids." *Naturwissenschaften* 92: 255–264.

Schomburgk, R. 1837. "On the identity of three supposed genera of orchidaceous epiphytes, in a letter to A. B. Lambert." *Transactions of the Linnean Society of London* 17: 552.

Slaughter, T. (ed.). 1996. *Bartram: Travels and Other Writings.* New York: Library of America.

Sousa-Baena, M. S., N. R. Sinha, J. Hernandes-Lopes, and L. G. Lohmann. 2018. "Convergent evolution and the diverse ontogenetic origins of tendrils in angiosperms." *Frontiers in Plant Science* 9: 403.

Sprengel, C. K. 1793. *Das Entdeckte Geheimniss der Natur in Bau und in der Befruchtung der Blumen.* Berlin: F. Vieweg.

Wallace, A. R. 1867. "Creation by law" [critical review of *The Reign of Law* by the Duke of Argyll, 1867]. *Quarterly Journal of Science* 4: 471–488.

Wedgwood, L. 1868. "Worms." *Gardeners' Chronicle and Agricultural Gazette* 19 (28 March): 324.

van Wyhe, J. (ed.) 2009. *Charles Darwin's Shorter Publications, 1829–1883.* Cambridge, UK: Cambridge University Press.

IMAGE CREDITS

All botanical illustrations provided by Oak Spring Garden Foundation, Upperville, Virginia.

All of Darwin's original woodcut illustrations provided by the Mertz Library of The New York Botanical Garden.

Page 15, Photo of Charles Darwin at age seventy-two on the veranda at Down House, with *Parthenocissus* twining up the post. Photo taken in 1881 by Messers. Elliot and Fry, published in *The Life and Letters of Charles Darwin* (1887).

Page 20, Down House, Darwin's home in Kent. Wood engraving by J. R. Brown, courtesy of the Wellcome Collection.

Page 24, Darwin's study at Down House, complete with experimental plants. Wood engraving by A. Haig, courtesy of the Wellcome Collection.

Page 26, Darwin in his Greenhouse. *Illustrated London News*, December 1887.

INDEX

experiments, 11, 14, 21–22, 25, 27,
  56–57, 129–131, 133–134, 179,
  186–187, 194, 217, 219–220,
  225–227, 230, 234, 240, 260,
  272–273, 279–280, 295–296, 328

F
Fairchild, Thomas, 109
fava bean, 314–321
Ferrari, Giovanni Battista, 344
fertilisation *See also* cross-fertilisation;
  pollination; self-fertilisation
  beans, 252–254
  Fragaria, 153
  Gloriosa, 158
  orchids, 23–24, 44–46, 69–75, 89–93,
    101–106, 146–149, 223–225
  Phaseolus, 251–255
  research, 31–32
"Fertilisation of British orchids by insect
  agency" (Darwin, C.), 224–225
"Fertilisation of Vincas" (Darwin, C,), 325
Fischer, Tom, 13
Fitch, John Nugent, 88, 350
Fitch, Walter Hood, 39, 294, 345, 349,
  350
Flagellaria indica, 158
flame lily, 156–161
flame nasturtium, 309
flax, 29, 56, 60–61, 190–195, 279
flies, 33, 105, 126, 132, 149, 258, 261
Flora Londinensis (Curtis), 344, 345
A Flora of the State of New York (Torrey),
  349–350
Florilegium Renovatum et Auctum (Bry),
  343–344
flowers, forms of
  Angraecum sesquipedale, 42–47
  botanical books, 16, 29
  Catasetum, 68–75
  Coryanthes, 88–93
  Cypripedium, 100–107
  Epipactis, 144–149
  Fragaria, 150–155
  Linum, 190–195
  Mitchella repens, 216–220
  Ophrys, 222–227
  Orchis, 228–235
  Oxalis, 237–240

Phaseolus, 250–256
Primula, 270–277
Pulmonaria, 280–283
research, 30
Salvia, 284–287
Spiranthes, 294–299
Vinca, 322–325
Viola, 328–330
Flowers, Moths, Butterflies and Shells (Ehret),
  345–346
Fly Orchis, 224
foxglove, 29, 114–118
Fragaria, 150–155
Fragaria vesca, 150, 152, 343
France, 152
Franken-orchid, 70
Franklin, Benjamin, 17
French pea, 264
French School artist, 322, 347
Fridericia mollissima, 55

G
Galápagos Islands, 19–20
Galton, Francis, 171
Gardeners' Chronicle (journal), 27, 29, 188,
  223, 224–225, 251–252, 324, 350
garden nasturtium, 309
garden pea, 262–269
Gartner, Karl Friedrich von, 25
Gauci, Maxime, 341
George Inn (George & Dragon), 163
geotropism, 49, 51, 98, 154, 155, 240,
  306, 320
Germany, 268
Giacometti, Diego, 340
Givenchy, Hubert de, 340
Gloriosa, 156–161
Gloriosa plantii, 158
Gloriosa simplex, 158
Gloriosa superba, 156, 343
Goessens, A., 68, 348
golden rain tree, 64
gourd, 138–143
Graham's sage, 286
Grandval-Justice, Sophie, 340
Grant, Robert Edmond, 17–18, 19–20
grape phylloxera, 333
grape vine, 332–337
grass vetchling, 177

gravitropism, 34
Gray, Asa, 10, 16, 22, 24, 27, 56, 61, 89,
    101–102, 130, 139, 141, 206, 217,
    219, 229, 257, 280, 334, 349
green bean, 251
groundnut, 49–53
Guayteca Island, 289

H
Haig, A., 24
Hamilton, Ann, 76, 168, 190, 208, 250,
    270, 347
Hansen, George, 349
Hardy, Charles, 304
heartsease, 327
heart-seed, 63
helleborine orchid, 144–149
Henslow, John Stevens, 10, 18, 19, 25,
    32, 109, 334
    arium, Royal Botanic Gardens, Kew, 27,
    122, 246, 323
Herbarius Ad Virum Delineatus (Withoos),
    351
Herbert, William, 264
Herbier Général de l'Amateur (Bessa), 343
heterostyly, 25, 32, 191, 217, 218, 237,
    272, 280
Hildebrand, Friedrich, 237, 281, 283,
    286
hive-bees, 149, 181–182, 254, 268, 303,
    304
HMS *Beagle*, 39, 209, 289
Holden, Samuel, 341
honeybees, 303
hook and root climbers, 33
Hooker, Joseph Dalton, 10, 19, 20, 21,
    22, 23, 27, 39, 43, 53, 64, 69, 77,
    83, 129, 140, 158, 198, 209, 220,
    230, 245, 301, 335, 345
Hooker, William Jackson, 39, 69, 284,
    345
hops, 162–167
horse-chestnut, 132, 259
Howard, Frances, 114, 128, 347–348
humble-bees, 89–93, 110, 111, 115, 254,
    283, 285–286, 299, 301
"Humble-bees" *Gardeners' Chronicle*
    (Darwin, C.), 285
Humboldt, Alexander von, 18, 25

Humulus lupulus, 162–167, 351
Huxley, Huxley, 273

I
incarnate clover, 303
insectivorous plants
    Drosera rotundifolia, 128–137
    experiments, 25, 33–34
    Pinguicula, 256–261
    sundew, 29, 33, 121, 128–137
    Venus fly trap, 34, 120–127
Insectivorous Plants (Darwin, C.), 29,
    123–127, 130, 131–137, 258–261
insect pollination, 31, 43–46, 70, 71–75,
    91–93, 102–106, 110, 111, 145–
    149, 158, 178–179, 181–182, 185,
    186, 188, 192, 194–195, 223–227,
    230–234, 252, 266–268, 286, 287,
    295–299, 302–303, 310, 315, 323,
    324, 325, 328
inter-fertility, 217, 280
International Dianthus Register,
    109
Ipomoea, 168–175, 178, 351
Ipomoea nil, 36
Ipomoea purpurea, 168, 347
Ips, 169
Italy, 179

J
Jadera haematoloma, 64
Jardin du Roi at Versailles, 345
John F. Kennedy Presidential Library,
    339
Johnny-jump-up, 327
Journal of Horticulture, 29
Journal of Researches (Darwin, C.), 289
Journal of the Linnean Society of London, 29,
    130
Juncus, 259

K
Keen's Seedlings, 153
Keller, Christopher, 350
Kennedy, John F., 339
Kent Wildlife Trust, 230
kidney bean, 252–254
Kilburn, William, 344
Koelreuteria elegans, 64